On the Eve of the Cybercultural Revolution

On the Eve of the Cybercultural Revolution

Black Power and Capitalism in the 1960s

Brian Bartell

University of Minnesota Press
Minneapolis
London

Chapter 2 was originally published as "The Political Ecology of James and Grace Lee Boggs," *Rethinking Marxism* 33, no. 3 (2021): 396–414, doi.org/10.1080/08935696.2021.1923184. Chapter 3 was originally published as "Noah Purifoy's Aesthetic for the Racial Capitalocene: Reading *66 Signs of Neon*," *Cultural Critique* 112 (2021): 24–58, dx.doi.org/10.1353/cul.2021.a790983. Chapter 4 was originally published as "The Material Imperialism of Home in Paule Marshall's *The Chosen Place, the Timeless People*," *Journal of Literary Studies* 36, no. 1 (2020): 25–42, doi.org/10.1080/02564718.2020.1738713.

Published by the University of Minnesota Press
111 Third Avenue South, Suite 290
Minneapolis, MN 55401-2520
http://www.upress.umn.edu

ISBN 978-1-5179-1392-2 (hc)
ISBN 978-1-5179-1393-9 (pb)

A Cataloging-in-Publication record for this book is available from the Library of Congress.

Printed in the United States of America on acid-free paper

UMP BmB 2025

Contents

Introduction

James Boggs began his 1964 speech "The Negro and Cybernation" by arguing, "To visualize the future rôle of Negroes in a cybercultural society, one must review, if only briefly, their past rôle in American society and what this means at the present stage of industrial development."[1] Much like today, where the news is replete with stories of automated unemployment and multivalent, technologically determined crises, the 1960s was also a time of widespread worry in the United States—worry about the ramifications of automation and the extent to which it seemed to be undermining the Keynesian management of capitalism in ways that programs like the Great Society and the War on Poverty were incapable of addressing. Also like today, in the 1960s these technologies were primarily presented as being new and unprecedented. However, as Boggs's remark demonstrates, he asked listeners instead to see the "present stage" as an historical one in which what was new was also connected to a longer political history of race, class, labor, and technology dating to plantation slavery. Throughout the period, James and his wife, Grace Lee Boggs, argued that this history conditioned the emergence of cybercultural society and that political movements for a classless society must be attuned to both this history but also to the ways that automation and cybernation required thinking beyond questions of employment and unemployment to racial and class composition, how technologies were connected to ecological relations, and what new forms of social reproduction and types of activities people must practice in order to create new values and to "initiate a new plateau, a new threshold on which human beings can continue to develop," which would be "the result both of long preparation and a profoundly new, a profoundly original beginning," as they wrote in *Revolution and Evolution in the Twentieth Century*.[2]

I begin with "The Negro and Cybernation" because James and Grace Lee Boggs's political theory is central to my examination of the period I am calling, following their formulation in "The City Is the Black Man's Land," the cybercultural era. *On the Eve of the Cybercultural Revolution: Black Power and Capitalism in the 1960s* is an historical study of Black political economy, in the broadest sense of the term, in which I will take on three tasks. First, I argue simply that theorizing the politics of automation in the 1960s, as well as affiliated terms like *mechanization* and *cybernation,* was central to a heterogeneous and capacious Black Power movement. An examination of Black Power cyberculture aids in both rethinking what constitutes that movement while also arguing that the cybercultural tendency provides an important approach to understanding the political histories of technocapitalism.

Second, through the Boggses' specific use of the concepts cybercultural era and cybercultural revolution, I offer a parallel history of the crisis of Keynesian state capitalism in the imperial United States. The history of the transition from Keynesianism to neoliberalism and/or Fordism to post-Fordism has been told many times, and my study shares much with influential versions of this history by, for example, David Harvey and, using a different set of concepts, Manning Marable's *How Capitalism Underdeveloped Black America.* Alternative to the period's popular use of the word *automation,* which, like today, named worries about technological unemployment that date to the beginning of industrial capitalism, or *cybernetics,* which referred either narrowly to computerized automation or to a broader field concerned with feedback and communication that traversed technology, society, and life in general, I argue that *cyberculture* was a conjunctural concept that named a specific mode of production and social reproduction. Black Power theorists understood the conjuncture to be emergent in the sense that automation was part of a political reorganization of employment and labor, the welfare state, race and class, and the dynamics of planetary capitalism and how it extracted value from human and other-than-human nature, which was undermining Keynesian forms of social reproduction. The cybercultural revolution thus opened two possible futures: it risked creating new forms of racialized inequality and dispossession through mass unemployment that would most profoundly affect already impoverished Black communities; and it offered new potential forms of freedom because automation meant that for the first time, Black peoples no longer had to necessarily labor for White capitalism.[3] The emergent technologies and political potentials of the cybercultural revolution were understood by Black Power, however, to be conditioned by the history of plantation slavery such that the cybercultural reorganization of capitalism was one moment in a history in which technological changes were always connected to hierarchical and racialized distributions of forms of labor, as well as forms of freedom and unfreedom, and more broadly how life was valued. In this respect, Black Power cyberculture's theorization of technocapitalism shows that the movement saw the cybercultural revolution as a specific conjuncture in what Saidiya Hartman calls the afterlife of slavery and had much in common with what Alys Eve Weinbaum calls a Black feminist philosophy of history, in that the plantation economy is sublated, both negated and preserved, in cybercultural capitalism, although here I examine a different set of issues than reproduction in contemporary neoliberalism.[4] Thus, although I do not offer a radically different telling of the crisis of Keynesian capitalism, I do hope that turning to Black Power cybercultural theory makes the conjuncture just strange enough to offer new perspectives on it, the Black Power movement, and the history of the automation discourse.

Third, the Boggses, Martin Luther King Jr., Noah Purifoy, Paule Marshall, the League of Revolutionary Black Workers (LRBW), and the Black Panthers, all subjects of individual chapters in this book, did not simply theorize technocapitalism but also offered differing political projects that sought alternatives to cybercultural inequality. In each chapter, I examine how they saw the mainstream Black freedom movement as being out of sync with the trajectories of technocapitalism. However, instead of offering utopian visions of automated liberation based on fetishizing automation as a production technology, their political praxis demonstrates that they treated it as a political-technological problem and sought alternative ways of organizing social reproduction in order to create a more-than-capitalist future attentive to the histories of racialized and gendered inequality and that sought new forms of value and human being in the world.

Throughout, I argue that Black Power cyberculture was theoretically and politically capacious. However, as my opening suggests, the conceptual approach is heavily influenced by the Boggses, and "The Negro and Cybernation" is a particularly rich distillation of their key concepts. The speech was delivered at the First Annual Conference on the Cybercultural Revolution, which took place in New York City between June 19 and 21, 1964, and would be collected in the 1966 proceedings volume that emerged out of the conference, *The Evolving Society.* This event was one of the most important conferences on what James Baldwin called "the cybernetics craze,"[5] and included, among other speakers, Hannah Arendt. The event was organized by Alice Mary Hilton, an electrical engineer and futurist who coined the term *cyberculture* and was one of the major, and now mostly neglected, figures of the period. The conference included presentations on a broad array of topics, including information processing and aesthetics, as well as asking whether "The Purpose of Business" was service or profit. Taking place against the backdrop of the Civil Rights Movement—including the recent publication of *The Triple Revolution,* a popular 1964 pamphlet; the *Technology and the American Economy* report by the National Commission on Technology, Automation, and Economic Progress; and the McCone Report on the 1965 Watts Rebellion, all of which I will discuss in the chapters to come—the conference acknowledged the extent to which racial inequality meant that the effects of automation were unequally distributed. However, the conference was generally an example of what Neda Atanasoski and Kalindi Vora term "techno-liberalism," where a universal human is understood to be the subject of the impacts of technological change in ways that obscure "the uneven racial and gendered relations of labor, power, and social relations that underlie the contemporary conditions of capitalist production."[6]

By the time the proceedings volume was published two years after the conference, in 1966, the political contexts of its reception had shifted. That was the year that Stokely Carmichael popularized the term *Black Power* during a "March Against Fear" speech in Mississippi. James and Grace Lee Boggs were among the earliest theorists of the heterogeneous and still highly debated term; James's own version of cyberculture foregrounds the historical approach to race, class, and capitalist forms of value that was a significant feature of Black Power.[7] Boggs argues that Black people in the United States were especially modern because of the entangled relationship between the production of race and the development of capitalist technologies, dating to the invention of the cotton gin during plantation slavery. In this respect, Boggs echoes C. L. R. James's claim in *The Black Jacobins* that the enslaved people of Haiti were the world's most modern at the time of the Haitian Revolution in 1791 because of the plantation's organization of collective labor, as well as recent claims by Louis Chude-Sokei that because of the simultaneous development of the categories of race and industrial technologies, "race" and "technology" have always had a mutually informing "creole" history.[8]

Boggs specifically argued that although Black people in the United States were primarily "scavengers" who had historically only "been entitled to the leavings, the castoffs of the whites," until the mid-1960s they had remained "inside" the capitalist economy.[9] Technological development before the cybercultural conjuncture had always meant an increased demand for more labor alongside machines, and thus more jobs for the Black working class, even if these were the worst-paid jobs available. However, "cybernation—*i.e.,* automation with nerve centers operated not by man but by computing machines—is eliminating the 'Negro jobs'" as the first step in the process of eliminating the vast majority of so-called productive labor.[10] However, thinking dialectically, he further contended that while this would limit working-class jobs, Black entry into "education" and "social and civil service jobs" was strategically important. While these were not necessarily the jobs that Black people "consciously" wanted—that is, "to be employed technically like whites"—these were in fact the types of work that "will become a major factor in determining the final disposition of the results of technology, *i.e.,* in the revolutionary and political arena where decisions will be made that govern the use of things rather than how they are to be produced. And these are the most important decisions in a cybercultural society."[11] In "The Negro and Cybernation," James Boggs politicized technological change. Automation and cybernation do not simply have racialized and classed effects; instead, race, class, technology, and capitalism have an evolving constitutive relationship. This historical and theoretical perspective grounded his contention that the Civil Rights Movement was out of sync with developments in technocapitalism and that integration into the Keynesian economy was increasingly impossible because industrial-based, family wage, jobs were being eliminated. This was a perspective that James Boggs shared with other Black Power cybercultural theorists, and each chapter will examine their different approaches to this problem.

Although James and Grace Lee Boggs posited the importance of education and social service jobs, he argued that these were primarily transitional forms into what they called elsewhere "socially necessary activity." As James noted, "A Job Ain't the Answer" to the inequalities and contradictions of capitalism, and the goal of their version of a countercybercultural revolution was a revolutionary reorganization of life beyond the commodity form.[12] As a new conjuncture in technocapitalism in which race and class were being recomposed and productive labor was, they believed, on the verge of being eliminated, they argued that a new political project, developed out of the histories of radical labor organizing and the Black freedom movement, must create new forms of value. To cite perhaps his most famous line from 1963's *The American Revolution:* "Now that man is being eliminated from the productive process, a new standard of value must be found. This can only be man's value as a human being."[13] More than any other, these lines animate Black Power cyberculture's political project. Capitalism relies on racialized distributions of waged and unwaged, productive and reproductive labor, which is connected to how capitalism relies on cheap nature and racialized distributions of waste in order to extract surplus value. Alternative to these, Black Power cyberculture sought new standards of value, whether through a more expansive welfare state, socialist and degrowth political economies, community control of technology, or new practices of creativity. Throughout, I take a conjunctural approach that examines how different Black Power activists, artists, and writers conceptualized both the politics of technocapitalism and the projects through which they sought new more-than-capitalist genres of the human. Today, what Aaron Benanav calls the automation discourse is resurgent.[14] While its specificities are different (including algorithmic racism, mass incarceration, debt, climate change, neoliberalism, health and reproductive crises, and the renewal of fascism), the cybercultural era is a key moment in the history of the present, and I return to the Black Power movement in the hopes that their politics might inform how racial capitalism, automation, and alternative futures are approached today. In the rest of the introduction, I elaborate on the necessarily interlinked key terms in the book's title: cyberculture, Black Power, capitalism, and revolution.

Cyberculture

For most readers, the term *cyberculture* likely evokes early internet theorizing of the 1990s and 2000s.[15] However, *cyberculture* is not a term from the 1990s. It was first used by Alice Mary Hilton in the 1960s and was derived from *cybernetic,* coined by Norbert Wiener in 1947, which he based on the Greek word *kybernetes,* meaning "to govern" or "to steer." During this period, Wiener, along with Margaret Mead, Gregory Bateson, and others, saw cybernetics as not only a way of considering how automated machines could work through increasingly advanced feedback mechanisms but also as an expansive concept that provided a way of seeing fundamental parallels between machines, human and nonhuman life, and social systems as forms that operated through communication and feedback. Although it failed to take hold as an academic discipline, its influence continues today in everything from machine learning to climate science.

The cybernetics craze of the 1950s and 1960s has increasingly become the subject of scholarship; however, the term *cyberculture* as it was initially defined by Hilton and then reworked by the Boggses has been largely forgotten.[16] Alice Mary Hilton deserves a book of her own, but here I am primarily interested in how James and Grace Lee Boggs drew on her concept to formulate a specifically Black Power cybercultural politics. Hilton defined cybercultural society broadly as one in which cybernation technologies based on feedback and communication offered the possibility of "the complete emancipation of human beings from drudgery and scarcity."[17] At the Evolving Society conference, Hilton argued that cyberculture should be understood in terms of three primary features: "first, the unprecedented rate of acceleration of change; second, the deceleration of the span between the conception of an idea and the invention of a technique; and third, the enormous contrast between our technological sophistication and our social-philosophical infantilism."[18] In today's language, Alice Mary Hilton could be considered an accelerationist—someone who argued not only that technology was rapidly developing but also, and more important, that that development should be embraced as a way to go through the contradictions of the technological present in order to arrive on the other, postscarcity, side. However, she also argued that cybercultural society's "social-philosophical" perspectives were out of sync with its technological present and futures. This is a perspective shared by the Boggses and especially by Martin Luther King Jr., whose cybercultural politics will be the subject of chapter 1.

The Evolving Society conference was organized to address, especially, the third point in light of the first two, because the technological break that cyberculture seemed to represent from industrial society was at a "magnitude," she argued, that exceeded the break between agricultural and hunter-gatherer societies—not only in its technological aspects but also in its impacts on humanity's "structures of feeling," to use Raymond Williams's still felicitous phrase.[19] This was in part due to the degree to which cybernetics opened up a way of seeing complex machine–human–environmental interrelationships, and because cybernated technologies were overproductive and creating, especially in the United States, an actual and potential situation in which surplus production was met by underconsumption—one of the period's dominant conceptualizations of the problem of automation. Thus, even in Hilton's formulation, cyberculture was a political-economic issue that presented the dual problem of managing increasingly complex machine systems and an economy where the techniques of Keynesianism seemed ill-equipped to manage the disconnect between production and consumption in the United States.

This also suggests the extent to which *cyberculture* was a periodizing term that in its dominant formulation ordered the past, present, and future of technology and humanity as a series of breaks. As Fredric Jameson has famously contended, the problem of periodization in the writing of history is inescapable. According to him, the choice between synchronic and especially diachronic "period formulations always secretly imply or project narratives or 'stories'—narrative representations—of the historical sequence in which such individual periods take their place and from which they derive their significance."[20] Like today, automation and cybernation were primarily seen as new issues, even if some, like Harry Braverman, understood automation, as a political/labor problem, to be a recurrent feature of capitalism.[21] Keeping the tension between what was novel about the historical moment and what was continuous with the past was essential to Black Power approaches. However, as I have begun to argue, they foregrounded a political-economic history derived specifically from plantation slavery and not simply a history of technological development, or that of an abstract, historically consistent, capitalism.

That Black Power cybercultural theorists would make the connection between automation and the afterlife of plantation slavery is perhaps unsurprising. However, Alice Mary Hilton's own sense of what made the cybercultural era distinct was also formulated in terms of the histories of slavery. Hilton ended her introduction to the proceedings, the "Bases of Cyberculture," by writing:

> The machines of this new age that can emancipate us and allow us to lead truly human lives, frighten those who are afraid of their own weakness and incompetence. To those who have no faith in man, the machines seem threatening to destroy mankind. But machines cannot destroy mankind; only we, our present generation of humanity, can destroy our own and future generations if we do not learn to be wise as well as omnipotent masters who can use cybernated slaves intelligently and for the good of mankind.[22]

Cyberculture was for Hilton a mode of social organization that offered previously unimaginable forms of power and emancipation from the most base, laboring, aspects of human existence, if managed properly. However, the flip side of this emancipatory potential, as in so much science fiction, are specters of human and machine enslavement. In light of James Boggs's own reflections in "The Negro and Cybernation," it is impossible not to read Hilton's phrasing in terms of the history of racial slavery in the Americas. Today, scholars like Neda Atanasoski, Jonathan Beller, Ruha Benjamin, Simone Browne, Wendy Chun, Kate Crawford, Louis Chude-Sokei, Curtis Marez, Alondra Nelson, Safiya Umoja Noble, Astra Taylor, Kalindi Vora, and Jackie Wang are increasingly examining the extent with which supposedly neutral technologies, like automation, produce unequal racialized and gendered effects; further, they are also historically conceptualized in terms of social difference, colonialism, and slavery.[23] Atanasoski and Vora argue that modernity's "automation fantasies" "(1) depend upon slavery and the idea of a worker who cannot rebel and (2) are contiguous with existing labor exploitation along colonial/racial lines." Further, the universalization of machines and humanity obscures the ways that both are part of racialized formations of freedom and what it means to be human.[24]

In the above passage, Hilton universalizes forms of enslavement and freedom precisely at a moment in which the inequalities of the afterlife of slavery were being openly contested in conjunction with Third World decolonization movements. James and Grace Lee Boggs define their sense of cybercultural revolution and the importance of Black Power politics at this specific moment most fully in the 1966 essay "The City Is the Black Man's Land."[25] Instead of universalizing the politics of automation and cybernation, here, as in "The Negro and Cybernation," they examine automation in terms of the history of White Power and capitalism, as well as the specific distribution of political-economic conflicts at that moment. I will return to this essay in greater detail in chapter 2; however, here I want to focus on four points that are important to how cyberculture will be used throughout the book, even if not everyone in the Black Power movement fully subscribed to them.

The first of these four points is that the primary political protagonists of the urban rebellions of the period are Black "outsiders" who have been rendered "socially unnecessary by the technological revolution of cybernation and automation," and who are the vanguard of a generalized condition to come even if the racial politics of that future are yet to be determined.[26]

Second, unlike in industrializing postcolonial countries, "North America has already completed this [industrial] revolution and is on the eve of the cybercultural revolution."[27] For the Boggses, cybercultural revolution thus marks an historical organization of capitalism, value, technology, labor, class, and race. Black Americans had already undergone modernization and were being, as they say elsewhere, "undeveloped" and rendered increasingly "socially unnecessary" by emergent mass automated unemployment, especially in industry.[28] The Boggses shared this view with others, like Eldridge Cleaver. However, this position was contested; alternatively, the LRBW argued that automation's most important feature was intensified Black labor and not necessarily unemployment. As this also suggests, the Boggses primarily focused on the "American Revolution," even if they placed it in a planetary frame. Yet the racial politics of cyberculture exceeded the boundaries of the U.S. nation-state, and in chapter 4, especially the discussion of Paule Marshall's novel *The Chosen Place, the Timeless People,* locates domestic cyberculture in relation to development, U.S. imperialism, and the plantation economy in the Caribbean.

Regardless of these differences, Black Power cyberculture generally argued that technocapitalism historically relied on racialized distributions of waged and unwaged labor, forms of dispossession, and corresponding distributions of freedom and unfreedom that were being articulated anew in the cybercultural era. The Boggses argued that because productive labor was becoming obsolete, new forms of "socially necessary activity" were becoming important, like education, care work, and social services.[29] Jason Smith argues that the future of high automated unemployment in the United States that the Boggses, Hilton, and so many others in the 1960s believed would happen did not take place because of the dramatic increase in service jobs that absorbed the lost manufacturing jobs, which were in part caused by automation.[30] Writing in 1966, the Boggses believed that instead of service "jobs," these were political activities that would "establish relations between man and man."[31] Most important, however, the Boggses' formulation in "The City Is the Black Man's Land" foregrounds that for them, cyberculture was characterized by both a new form of production and new forms of social reproduction beyond Keynesianism. It was especially new forms of social reproduction, not new forms of employment or technological fetishism, that would be the focus of their politics, and it was through this lens that they reconceptualized new valuations of the human beyond capitalism.

Finally, the Boggses argued that because of the historical trajectory of capitalist dispossession, Black people must necessarily be at the forefront of political revolution. For them, this would take place in U.S. cities that were now majority Black—sites where all of the political, economic, and environmental problems were congealed. However, this would also be a planetary project: by "Black," they meant "all people of color who were involved in revolutionary struggle."[32] Thus, they argued that Black Power was not racist, as it was already being called at this early moment. Instead, it was a political project that was partly about immediate Black urban political power, but it was also a much more expansive project to create a classless society.[33] While the Boggses' sense of cyberculture provides the frame for thinking about Black Power's engagement with the politics of technocapitalism in the 1960s, the movement was far from monolithic. In the pages to come I hope to establish Black Power cyberculture as a coherent theory of technocapitalism—but also one that is capacious and open to debate and disagreement, thus giving a fuller sense of its dynamics and political potentials.

Black Power

As I have begun to show, James and Grace Lee Boggs's formulation of Black Power in "The City Is the Black Man's Land" defined it as a mode of political economic analysis focused on understanding the emergent features of capitalism and anticapitalism in terms of the dialectical history of race, class, labor, nature, technology, and value. As I discuss more fully in chapter 2, they called this approach "scientific" Black Power, drawing on the history of Marxist thought. Implicit in the discussion in "The City Is the Black Man's Land" of the founding in 1965 of the Organization for Black Power in Detroit is that far from something that is added on to an already existing, stably defined Black Power, cyberculture was constitutive of Black Power's emergence and therefore was a central feature of it. As Jackie Wang argues, Black Marxist theorizations of automation have largely been ignored, and in *On the Eve of the Cybercultural Revolution,* I seek to expand how Black Power is defined and also to insist that the movement's thought is important for understanding the politics of automation during its own time as well as today.[34]

The histories of Black Power date at least as far back as the communist Black Belt theories of the 1920s. *Black Power* was used in different, if complementary, senses by Richard Wright and Harold Cruse before the term's full emergence in the 1960s. However, popularly, the movement is associated with the 1960s, and in the years of counterrevolution that have followed, it has largely been reduced to a narrative in which a separatist, militant Black Power movement replaced the interracial and integrationist Civil Rights Movement around 1965. This reductive view includes a few ideas about Black "cultural nationalism" and "self-defense" and images of Malcolm X,

Angela Davis, the Black Panthers, and a few others in which their politics are largely evacuated in favor of fashion.[35] Over the past fifteen years, scholarship by Ashley Farmer, Cedric Johnson, and Peniel Joseph, along with many others, has begun to shift the focus of what Joseph has described as the new "Black Power Studies," expanding the periodization of the movement and redefining it as a "complex mosaic" of ideas, actors, and geographies, including, but not limited to, debates about what constituted a "Black Aesthetic," the significance of Black feminist activists, the role of internationalists and an interracial Third World politics, and lively discussions about the different roles of urban versus rural communities.[36]

Here I focus on a much narrower segment of the movement than Joseph's edited volume, *The Black Power Movement: Rethinking the Civil Rights–Black Power Era,* or Cedric Johnson's *Revolutionaries to Race Leaders: Black Power and the Making of African American Politics.*[37] However, that complex mosaic of ideological debates is essential to how Black Power cyberculture was theorized and practiced, whether this be debates over gendered labor, the family, and the politics of the welfare state; those over domestic colony theory and whether nationalist, internationalist, or intercommunalist approaches were the most incisive ways of understanding technocapitalism; those regarding who the vanguard of the cybercultural revolution should be; those about the role of cultural nationalism, class, and gender as they relate to how capitalism mobilizes race; and those about whether automation was even happening or if instead it was just another example of capitalism's ongoing racialized distribution of labor. Alongside these explicitly political debates, I am also concerned with the formal practices that activists, artists, and writers used to theorize cyberculture. This is true of Martin Luther King Jr.'s speeches, James and Grace Lee Boggs's low theory, the Black Panthers' use of print culture, Noah Purifoy's junk assemblage aesthetic, Paule Marshall's realist novel *The Chosen Place, the Timeless People,* and the LRBW's movie *Finally Got the News* as well as their editorial practices more broadly. The formal features of the Black Aesthetic were much debated in the period, though in general the goal of the Black Arts Movement was to create artistic and critical practices that were autonomous from normative, White, aesthetic values, exhibiting institutions, and power dynamics.[38] However, I am less interested in making overt claims about a

Black aesthetic than I am in treating these as works of theory that use their different aesthetic forms to understand the dynamics of cyberculture and that themselves can be thought of as practices of social reproduction through which alternative forms of value and knowing are brought into being.

My approach to Black Power thus has two primary aims: it is an historical study that seeks to demonstrate the importance of cyberculture to the movement while also being a work of Black Power cybercultural theory. In this sense, it follows Brandon Terry's contention that instead of defining Black Power, we should draw on David Scott's ideas to consider the "problem-space," or ensemble of questions and answers, through which the political horizons of Black Power were oriented.[39] At the center of this problem-space was the technocapitalist organization of life in the 1960s as well as its histories and potential futures. The political problem of technology to the Black freedom struggle was far more expansive than I have space to fully discuss in a single book.[40] By focusing on the work of Martin Luther King Jr., Grace Lee and James Boggs, Noah Purifoy, Paule Marshall, and the LRBW, and under the umbrella of the Black Panthers, Eldridge Cleaver, Elaine Brown, Angela Davis, George Jackson, and Huey P. Newton, I explicate how automation and cybernation brought to the fore a history in which racialization and technology change, beginning with plantation slavery and moving through agricultural mechanization, had always been interconnected, while also raising the question of what the politics of Black Power look like in terms of this history. In contrast, however, to critiques by comrades of the Black freedom movement like Raya Dunayevskaya, who argued in 1967 that Black Power wasn't attentive enough to working-class struggles and interracial alliances, I examine how Black Power cyberculture understood race *and* class to be necessarily intertwined while continually being recomposed in the political struggles over the organization of life by and against technocapitalism.[41]

The invocation of class tends to evoke images of the shop floor and the point of production, and these will certainly be present in the LRBW, as well as in work by Paule Marshall and the Boggses. However, Black Power cyberculture formulated class far more expansively. An examination of automation and its capacity to create mass un- and underemployment calls into question conceptualizations of labor and surplus-value extraction, and thus how class was being recomposed in and beyond the workplace. Moving away from assumptions about productivity, gendered labor, and the family wage, I argue that Black Power cyberculture began to theorize the ways in which race, gender, class, and the consumer were categories contingent on normative models of organizing labor power and labor power's reproduction whose boundaries were unstable—whether this be the status of the lumpenproletariat, the worker, the "outsider," or the community, as well as those categories' gendered and racial boundaries; how reworking the "junk" produced by space-age technocapitalism might be reworked and revalued by those working in the informal economy in its shadows; or how domestic spaces in the Caribbean were not only sites that reproduced hierarchies from slavery in a U.S.-dominated neocolonial plantation economy but were also spaces of speculation about the connection between cybernetics, gender, race, sexuality, and changes in planetary capitalism.

Famously in the period, Martin Luther King Jr. called for a move from a "thing-oriented" to a "person-oriented" economy—a demand that he shared, though often in very different language, with other Black Power cybercultural theorists. Reading them together demonstrates that this call also demands that technocapitalism be understood in terms of the total process of value extraction in and against which race, class, and gender were (re)composed. This begins to suggest the extent to which Black Power cyberculture endeavored to theorize how capitalism exploited, produced, and organized forms of social difference in order to not only understand these processes but ultimately to create new forms of value—an undertaking that required reworking old categories in new contexts and creating new ones entirely. The LRBW defined Black Power as Black labor power, a definition that I argue was based on their expanded theory of production, and that was of utmost importance at a moment when automation was undermining the value of labor power. Following the Boggses, however, my primary focus will be on the extent to which Black Power was not so much Black labor power but instead Black *activity*—an ensemble of intellectual, political, working, caring, and artistic practices out of which, to borrow from Sylvia Wynter, a new human praxis might emerge. At its most expansive, Black Power cyberculture would argue that a new human praxis required an engagement with the ways that capitalism puts life to work in general, which would act as the basis for new political forms and practices that would permit the creation of a more-than-capitalist world that justly distributed the surplus produced by all.

Capitalism

The initial motivation for *On the Eve of the Cybercultural Revolution* came from the question of value as it appeared in James Boggs's book *The American Revolution: Pages from a Negro Worker's Notebook* and in Noah Purifoy's essay "The Art of Communication as a Creative Act," which was included in the catalog to the post–Watts Rebellion junk assemblage art exhibition *66 Signs of Neon.* I have already quoted above Boggs's claim that the political-economic conjuncture demanded a new and human standard of value. Purifoy stated in his essay that "Watts finds itself virtually set down in the center of junk piled high on all sides. Its main industry is junk! The essential question being posed to the community by the exhibit was, what is the true value of these materials over and above sale to junkyards for a few cents?"[42] On the surface, they appear to be quite different, with the first about the industrial production of commodities and the second about what is left over after they have been used up and junked. Together, they present a picture of the capitalist valorization process as it moves from extraction through production to consumption and then disposal and, potentially, recycling, all of which require differential forms of racialized labor. What they ultimately ask is how new forms of self-valorization and thus new genres of the human can emerge out of capitalist relations of value.

From the outset, I have argued that cyberculture names a mode of production and social reproduction in which the plantation economy is sublated, but that was also seen by Black Power theorists as a moment in which race, class, labor, and value were being recomposed by capitalism's deployment of new automation and cybernation technologies. As I also note, through the term *cyberculture,* I hope to estrange a political-economic conjuncture that it has become standard to describe as the end of Keynesianism and the beginning of neoliberalism. In this sense it shares something with David Harvey's *The Condition of Postmodernity* when the term *neoliberalism* was, for him at least, unavailable to describe the politics and systemic contradictions of that moment in which "the success of Fordist rationalization meant the relative displacement of more and more workers from manufacturing" and where Keynesian welfare statism (and its organization of social reproduction) was no longer viable, especially as its inequalities were challenged on multiple fronts.[43] Given that I argue that what was at stake for Black Power was theorizing the present and history of value extraction as the first step in a political praxis to create alternative forms of value, I hope to avoid reducing their interventions to a teleology of neoliberalism, even if, as an historical study, this is largely unavoidable, and by doing so open them to a new future to come. In the title, however, I have chosen to retain the unmodified term *capitalism* because of the two primary traditions of political economic thought that inform the book: racial capitalism and social reproduction theory.

The term *racial capitalism* became widely known through Cedric Robinson's *Black Marxism: The Making of the Black Radical Tradition,* which borrowed a term developed by South African antiapartheid activists to describe the extent to which hierarchically racialized social difference is central to capital accumulation strategies. Although it has been taken up by Marxist theorists like Ruth Wilson Gilmore, Charisse Burden-Stelly, and David Roediger, Robinson initially used it to critique the limits of then-current Marxist theorizations that reduced racial difference to a nonconstitutive feature of capitalism, one secondary to class that thus too often excluded direct confrontations with racial difference in anticapitalist horizons of freedom. Alongside this, in the first half of *Black Marxism,* Robinson retold the history of capitalism's emergence, as well as the history of the emergence of the category of race. While race is often convincingly dated to plantation slavery and to, for example, the institution of "partus sequitur ventrem" in seventeenth-century colonial Virginia, Robinson argued that what he called "racialism" preceded the enslavement of Africans.[44] As he wrote, "The tendency of European civilization through capitalism was not to homogenize but to differentiate—to exaggerate regional, subcultural, dialectical differences into 'racial' ones."[45] Instead of being a secondary feature of capitalism, racialism was one of the necessary conditions of the capitalist organization of labor power and accumulation generally. There is thus no capitalism that is not racial capitalism.

Theories of racial capitalism have now far exceeded Robinson's initial deployment of the concept to include gender, sexuality, other racial formations, geographies, and periods beyond the Black Atlantic, primarily masculine, context of *Black Marxism*.[46] Particularly important to my study is the extent to which racial capitalist theorization has opened up the question of capitalist periodization and its often linear emplotment, which I discussed above in relation to cyberculture.[47] Nikhil Pal Singh argues that alongside this rethinking of the temporalities of anticapitalist struggle, racial capitalist theorization has always been an especially conjunctural approach, emerging explicitly in Robinson and in Stuart Hall and others, out of the revolutionary and counterrevolutionary struggles and reconfigurations of capitalism in the 1960s and 1970s. This is also true of early theorizations of the relationship between race and capital accumulation before the introduction of the term *racial capitalism*—by, for example, W. E. B. Du Bois, Oliver Cromwell Cox, and Claudia Jones.[48] Framed thus, it should come as no surprise that Cedric Robinson was active in groups alongside Black Power cybercultural theorists like Huey P. Newton and Bobby Seale.

This approach to racial capitalism attempts to address, as Singh calls it, the "durability" of racism and racial difference, even, or especially, at moments of Black inclusion, like into wage labor after the Civil War, and into the Fordist–Keynesian family wage in the cybercultural era.[49] Black Power cybercultural theorists explicitly struggled with these issues and the results of that struggle, long underacknowledged, was a mode of theorizing that saw cybernetic and automated capitalism as an organization of relations of production and social reproduction that seemed to be renewing a number of historic tendencies. It was devaluing Black labor at time of struggle to become incorporated into Keynesian capitalism while simultaneously demanding of those employed in, for example, the automobile industry that they labor ever harder in order to make the so-called automated factory function. All this occured while increasingly distributing Black people in geographic proximity to the junk and environmental damage of Fordist overproduction. Yet the technological conjuncture was also introducing an apparently new dynamic into the history of racialism. If Black labor would no longer be central to surplus-value extraction, what did that do to notions of race and class? Would it mean that because they were at the avant-garde of the cybercultural revolution's production of outsiders that race would eventually be eliminated as a category of difference altogether in favor of new forms of class differentiation after White and other people of color became technologically unemployed? Or would it only be Black, and especially Black working-class and under- and unemployed, people who would remain outsiders in cybernated capitalism, thus creating new forms of racialized exclusion? This was the political problem that animated much of Black Power cybercultural organizing to create a more-than-capitalist world. Ruth Wilson Gilmore, Jackie Wang, Melinda Cooper, Jason Smith, and others have shown how forms of debt, incarceration, service work, and a returning emphasis on the family and gendered labor were the primary means by which neoliberalism and neoconservatism responded to the countercultural movements of the cybercultural era. However, returning estranged by way of Black Power cyberculture's approach to these problems offers new ways of seeing the period and demonstrates why Black Power should still be part of the critique of technocapitalism today.

Although my subtitle only uses *capitalism,* when appropriate, I will use both *capitalism* and *racial capitalism,* sometimes as synonyms and sometimes in order to emphasize the specifically racial dynamics at work in a given instance. Alongside racial capitalist theory, social reproduction theory is central to the book because, as I have already begun to show, while automation is almost always defined as a production technology that affects labor power and raises problems related to unemployment, Black Power cyberculture saw this as only a part of the problem. Automation raised a number of questions about how capitalism would reproduce itself after Keynesian mechanisms to ensure effective demand in the face of mass technological unemployment were exceeded. Most important, they theorized what new forms of social reproduction were desirable for a more-than-capitalist world and attempted to practice them in the present, of which the Black Panthers' survival programs are the best known.

At the most basic level, social reproduction theorists, drawing on and transforming Karl Marx's ideas as well as the work of theorists like Rosa Luxemburg, show the extent to which capitalism as a political-economic system, premised on the commodity form labor power to produce surplus value, legitimizes certain forms of labor as waged labor, while largely, and necessarily, relying on under- and unwaged forms of reproducing labor power, which are structurally separated and delegitimized from and by the so-called economic world. Forms of social reproduction include gendered domestic work in the family, as well as education, health, public services and state institutions, care work, and the work done by other-than-human life. As Nancy Fraser, Tithi Bhattacharya, and others argue, this productive/reproductive split has limited analytical approaches to understanding how capitalism's accumulation strategies and crises are driven by boundary struggles over production and reproduction, but perhaps most important, it has limited how anticapitalist political struggles are conceptualized by often legitimizing workplace struggles and delegitimizing other forms of social struggle as being noneconomic, which in turn limits how class formation is understood as well as its relationship to race, gender, and sexuality, and the struggles of minoritized peoples.[50]

Black feminist scholars like Ruth Wilson Gilmore, Saidiya Hartman, and Jennifer L. Morgan have been at the forefront of developing both social reproduction and racial capitalism as overlapping forms of theorizing how capitalism works, and they inform the pages to come.[51] The final chapter examines one of the formative texts of both racial capitalist and social reproduction theory, Angela Davis's *Women, Race, and Class,* especially the section "The Approaching Obsolescence of Housework: A Working Class Perspective." In it, Davis examines the extent to which it is possible to "socialize" housework and incorporate it "into the industrial economy" using "advanced cleaning machinery," thus destroying the domestic economy of un- and underpaid gendered and racialized labor and its "private" character.[52] This realizable potential is the starting place for an examination of the history of the division between productive and reproductive labor in the United States. In the context of political debates by the Black Panthers, in chapter 6 I am especially interested in her argument that Black workers should be central to the party's political strategy in part because of overly romanticized and gendered features of the lumpenproletariat. More generally, Davis's thought is central to Black Power cybercultural theory because she uses the technologies of housework to consider a history of social reproduction in the United States that is gendered, racialized, and classed in specific ways, then uses that history to examine potential politics in which not just labor but also the surplus produced by society might be organized otherwise.

Black Power cyberculture as an ensemble with individual specificities thus sought to theorize racial capitalism, primarily in the United States but also as a planetary system, as a mode of both production and social reproduction, and one "on the eve of the cybercultural revolution." In doing so, they sought to rethink gendered assumptions about waged and unwaged labor, ecological politics, educational practices, and forms of creativity and activity. They considered not only who would take on these political practices but also where and what forms this revolutionary activity should take. The cybercultural era was thus understood as a transitional moment in the extraction of value, or what Huey Newton called the ways that technocapitalism sought to transform life into a "reservoir of information" to be used for a further expansion of capitalist social relations.[53] In response, Black Power cyberculture's ultimate goal was to create, through new forms of socially necessary activity, an alternative, more-than-capitalist world. How they would do that takes us to the last term in this book's title: revolution.

Revolution

At the beginning of *Chronophobia: On Time in the Art of the 1960s,* Pamela M. Lee writes that "revolution is an unavoidable trope in the sixties historical record, a cliché even; and however we treat that revolution with hindsight—whether failed or hopelessly romantic or marginally successful—the vision of a time radically changed remains with us. Revolution, however, not only suggests a confrontation with authority but a peculiar mode of temporality."[54] Despite the often strident tones of the period, revolutionary time in the 1960s was, according to Lee, "troubled and undecidable" and was a figure for "uncertainty about the mechanics of historical change itself."[55]

As throughout the counterculture, *revolution* appears everywhere in Black Power discourses, and besides the *cybercultural revolution,* in the pages to come, it appears in many forms. These include *The Triple Revolution,* the popular 1964 pamphlet about automation and cybernation, human rights, and nuclear weapons revolutions that James Boggs coauthored and that Martin Luther King Jr. was in dialogue with. The League of Revolutionary Black Workers, of course, defined themselves as a revolutionary movement to overcome capitalism beginning on the assembly lines of Detroit and with the intention of ultimately taking state power. Of all of the political theorists of the 1960s, Black Power or otherwise, Grace Lee and James Boggs and Huey P. Newton were undoubtedly among the most thoughtful, nuanced, and sustained meditators on what revolution at that particular conjuncture actually meant.

Newton has become something of a stereotype of the 1960s revolutionary, especially in the famous image in which he sits, clad in a black beret and leather jacket, holding a rifle in one hand and a spear in the other, surrounded by African accoutrements, while seemingly meeting, and challenging, the spectator's gaze. However, during his 1971 discussion at Yale University with psychoanalyst Erik Erikson published under the title *In Search of Common Ground*, Newton presented a different figure, wearing "steel-rimmed spectacles" and speaking theoretically for "two hours in his thin, controlled voice without once using expressions like 'pig' or once employing any ornamental rhetoric."[56] Complementing this mode of presentation, Newton evidenced a great deal of uncertainty about revolution as well as his own status as a revolutionary. Beyond this persona, the undecidability of revolution is central to Newton's theorization of the politics of the conjuncture. This is the moment when Newton introduced his theory of intercommunalism, where he argued that because of technocapitalist "empire," the nation-state was no longer the dominant form of sovereignty and what he called "communities" would now be the sociopolitical form that must not only wage emancipatory struggles but also absorb and manage the impacts of capitalism's new forms of accumulation through dispossession at the ends of the Keynesian welfare state. What this meant, in part, was that the Black Panthers' theories of revolutionary nationalism and internationalism no longer held meaning; instead, they must organize a new politics on this emergent, uncertain terrain. But this was only one problem. Despite radical culture's belief that revolution would happen in one fell swoop and lead to a society free of contradictions, Newton said that as a "dialectical materialist," he believed that change driven by contradiction was at the center of not just political life but life in general. This meant that "anything can be revolutionary at a particular point in time."[57] He went further: "We can be very sure that there will be contradictions after revolutionary intercommunalism is the order of the day. . . . There will always be contradictions or else everything would stop. So it's not a question of 'when the revolution comes': the revolution is always going on. It's not a question of 'when the revolution is going to be': the revolution is going on every day, every minute, because the new is always struggling against the old for dominance."[58] Despite the desires and intentions one has for specific revolutionary outcomes, and certainly some form of revolution is necessary to create a more-than-capitalist society of equality, revolution is, according to Newton, an ongoing and indeterminate process. Revolution is, to borrow from Stuart Hall, without guarantees.

To reiterate, in "The City Is the Black Man's Land," the cybercultural revolution that the United States was on the eve of was not an emancipatory revolution but instead a technocapitalist one that, left to its own devices, would create a society of technological abundance but also one of mass technological unemployment, organized scarcity, and multivalent social reproductive crises. Writing in 1963, James Boggs called this political situation and the struggle against it "the American Revolution." What made the United States unique was, in part, the ways that race had been historically produced and exploited to divide Americans, but the problem of the American Revolution was especially that the United States was the "citadel of world capitalism" and a place with an "abundance of commodities."[59] Breaking with received ideas, he argued that revolution did not have to start with economic problems or the working class because U.S. capitalism had "generated more than enough contradictions to pose the question of the total social reorganization of the country."[60] These pages are suffused with not only the emergence of the Civil Rights Movement and the decline of radical labor but also with the fear and policing of the Cold War and the aftermath of McCarthyism. In this contradictory situation of abundance, inequality, and a society at odds with itself, Boggs argued that what was necessary to create a classless society were forms of "creativity" and "imagination" on par with those that had been put into "production."[61]

Grace Lee and James Boggs continued this line of thought in

what is arguably their magnum opus, *Revolution and Evolution in the Twentieth Century.* In it, they drew on the history of modern revolutions (the French, Soviet, Chinese, Vietnamese, and American ones especially), but what they argued above all was that revolution was not an instant or overnight process, as the mass media and so many 1960s-era "revolutionaries" claimed. Instead, using expansive language different from so much of their explicitly Marxist theorizing, they argue that it would be a long process that begins with "a projection of man/woman into the future. It begins with projecting a more human human being," one who is "creative," who is "conscious and self-conscious," and who has a "sense of political and social responsibility."[62] The people of the United States and the planet, they argued, were at a "threshold, a border, a frontier" in not only technological and material abundance but also environmental damage and organized scarcity. But there was also the potential for new values and for "initiat[ing] a new plateau, a new threshold on which human beings can continue to develop," although one "still situated on the continuous line between past and future."[63] The Boggses, especially during the cybercultural era, were vanguardists who believed that the role of the revolutionary was to be simultaneously student, visionary, and organizer—someone who learned from past revolutions, drew on what is best in the present, and channeled that energy through collective action into the long revolutionary process.

The Boggses' vision was not necessarily the vision of Black Power cyberculture in general. One thread that connects them all, however, is that the new society, and the political strategies that will bring it about, was not envisioned as one in which the machinery of automation and cybernation is just taken up and used unchanged, thereby somehow keeping the same relations of production and social reproduction while creating a society of equality. Instead, an alternative cybercultural society was envisioned, one attentive to the ways that those technologies emerged through histories of hierarchical exploitation, production, and the recomposition of race, gender, and class, as well as ways of making and doing that included waged and unwaged labor and other forms of activity. The revolution would thus not be a technological one in the sense of machinery but rather would be a revolution in values—in what people do, the concepts and categories through which people think and live, and ultimately in the distribution of social surplus.

In examining revolution, I begin with Martin Luther King Jr.'s movement for an expanded welfare state, as Black Power was replacing civil rights as the dominant language through which the politics of race was articulated, whether that be Black Power cyberculture, Black capitalism, or Black nationalism. I end with Huey P. Newton and the Black Panthers' belief that despite the desirability of a nation-state capable of securing equality through a universal basic income, universal health care, cooperative ownership of the means of production, and other similar programs, technocapitalism as a planetary organization of life for surplus-value extraction at the ends of Keynesianism seemed to make that impossible. In between are struggles over urban land, imperialism and development, the point of production, and artistic practices that attempted to meet the demands of cybercultural inequality. Today's automated moment is defined by different technologies and political-economic contradictions. I hope, however, that analyzing the expansive Black Power response to the cybercultural revolution will show how this thought is on a continuous line between past, present, and future.

1

"Remaining Awake Through a Great Revolution"

MARTIN LUTHER KING JR. AND THE TRIPLE REVOLUTION

On March 31, 1968, just five days before his assassination, Martin Luther King Jr. delivered his final Sunday sermon on the subject of "Remaining Awake Through a Great Revolution" at the National Cathedral in Washington, D.C. The speech condensed and developed his central ethical and political concerns from that period: his critique of the United States' war in Vietnam, the "network of mutuality" that connected what he was now calling the human rights (instead of civil rights) revolution in the United States to the global decolonization movements, and the generalized crisis of values facing the country. Most explicitly, the sermon was a moral call and justification for the upcoming, and controversial, Poor People's Campaign before religious and political leaders in the U.S. capital, which was to soon be the location of Resurrection City and the "massive" displays of civil disobedience to end poverty that King, the Southern Christian Leadership Coalition, and other groups were planning. "Yes, it will be a Poor People's Campaign. This is the question facing America. Ultimately a great nation is a compassionate nation. America has not met its obligations and its responsibilities to the poor."[1]

King believed this moment was defined by "a triple revolution": "a technological revolution, with the impact of automation and cybernation; then there is a revolution in weaponry, with the emergence of atomic and nuclear weapons of warfare; then there is a human rights revolution, with the freedom explosion that is taking place all over the world." Here King references the work of the Ad Hoc Committee on the Triple Revolution (AHC), a group of academics, intellectuals, and student and labor activists that included James Boggs, Michael Harrington, Tom Hayden, Brigadier General Hugh B. Hester, Alice Mary Hilton, Irving Howe, Gunnar Myrdal, Bayard Rustin, and Robert Theobald. The AHC was organized by a Santa Barbara, California–based think tank, the Center for the Study of Democratic Institutions. Four years earlier, in 1964, the AHC had authored a memorandum, *The Triple Revolution*, in response to Lyndon Johnson's Great Society–related War on Poverty program, which they thought insufficiently considered the changes being ushered in by new automation technologies.[2] *The Triple Revolution* was much discussed in publications ranging from *Life* to *Monthly Review*, and alongside the National Commission on Technology, Automation, and Economic Progress, it influenced debates in the United States over the extent and pace of technological change, the impacts of automation on labor and society at large, and whether some form of guaranteed income should be instituted to mitigate its effects.

While human rights and weaponry were of profound importance, *The Triple Revolution* focused exclusively on the automation and cybernation revolution, which, its authors argued, was ushering in "a new era of production" in which the "principles of organization are as different from those of the industrial era as those of the industrial era were different from the agricultural."[3] King follows the AHC in foregrounding the importance of automation in his sermon. Largely unlike *The Triple Revolution*, automation for King did not mark a technologically determinist break with past modes of production but rather marked an historical-political phenomenon. Automation was in fact reorganizing the Keynesian capitalist economy in the United States, "making all things new," King said, quoting the Book of Revelation, and calling into question normative conceptions of labor, the wage, and what it meant to be human. However, this revolutionary newness was not linear or detached from the histories of White Power in the United States. King argued in the speech that the triple revolution was part of a contemporary struggle between the "forces of goodwill" and "the extreme rightists" who were using "time much more effectively" than the human rights movement. He also argued throughout the 1950s and 1960s that the cybercultural revolution was conditioned by the history of capitalism, which, dating to the "plantation machine," had exploited and produced hierarchical, racialized forms of social difference. Much like Grace Lee and James Boggs in "The City Is the Black Man's Land," King saw a close connection between automation, the history of farm mechanization, and the forms of value extraction, dispossession, and un- and underemployment manifest in Black urban communities. In making these arguments, he drew on the work of W. E. B. Du Bois, especially "domestic colony" theory. In a speech at the Chicago Freedom Festival on March 12, 1966, he called this racialized organization of exploitation and dispossession the "General Economy of the slum."[4]

By examining Martin Luther King Jr.'s cybercultural theory, especially through his engagement with *The Triple Revolution,* this chapter argues that the politics of automation were central to the development of King's own political praxis and thus to the politics of Black Power more broadly. It also begins to define the cybercultural conjuncture as one in which automation was not simply a problem of primarily White male unemployment but a broader deregulation of Keynesian state capitalism's forms of social reproduction, thus opening a political struggle over how the economy should be reorganized. More specifically, I begin by examining *The Triple Revolution,* as well as a report, *Technology and the American Economy,* by the National Commission on Technology, Automation, and Economic Progress, to consider how they framed automation as an emergent problem that reflected the technological potential of Fordist rationalization, but one that, because of that success, was creating a crisis in the strategies through which Keynesianism ensured effective demand and social stability. At its most expansive, *The Triple Revolution* suggested that an entirely "new science of political economy" would be necessary in order to meet a new reality in which a universal human was no longer necessary to do productive labor.[5] However, most immediately, both documents suggested that some version of a guaranteed income was necessary to ensure the stability of consumer capitalism in a period of anticipated mass technological unemployment. As I discuss, a guaranteed income was an almost universally recognized need in the 1960s, and its implementation was promoted by King and the National Welfare Rights Organization as well as by Richard Nixon and Milton Friedman.

After this, I turn to Martin Luther King Jr.'s own engagement with the automation discourse. King discussed automation in his earliest public speeches, and it would become increasingly central to his critique of capitalism in the 1960s. Throughout, automation, although in many respects new, was for King also continuous with forms of dispossession that dated to the plantation machine in the United States and (settler and neo-) colonialism more generally. Initially, the political problem of automation was Black men's unemployment and the need to secure a family wage and access to "affluent society" for Black families—a project that, as Melinda Cooper argues, had broad appeal across party lines.[6] However, automation's capacity to intensify dispossession through unemployment began to push King to reconsider normative gendered assumptions about labor, employment, and the wage, as well as broader questions about how technocapitalism organized life to extract value and the extent to which it was necessary to reorganize a capitalist economy of "things" into a "people"-oriented one. To do this, I first examine King's general economy of the slum, and through it how he connected automation to the afterlife of slavery, domestic colony theory, and the emergent politics of Black Power. I then turn to his encounter with the Black women activists in the National Welfare Rights Organization and their argument for a guaranteed income disarticulated from normative models of the family and labor, as well as to his mobilization for the Poor People's Campaign, in order to consider what an expansive welfare state—one attentive to automation as well as the histories of racialized and gendered work and the geographies of inequality—might look like.

In these respects, King's thought offers an especially useful introduction to the development of Black Power cyberculture and a way of examining the possibilities and limitations of the welfare state in the 1960s. If the Keynesian welfare state was being deregulated through both technological change as well as challenges to its forms of exclusion by multiple counterculture movements, then to what extent could a reimagined welfare state reckon with the continued inequalities of the afterlife of slavery, and by so doing secure a more just distribution of society's surplus? This vision of a cybercultural welfare state was central to King's conception of the Poor People's Campaign and the human rights revolution more generally. Examining how King approached them will set up the chapters to come, permitting a discussion of the different ways that Black

Power cyberculture thought about welfare, more-than-capitalist forms of value, and the political practices through which they could be secured.

In examining King's cybercultural thought, this chapter takes up Tommie Shelby and Brandon M. Terry's challenge in *To Shape a New World: Essays on the Political Philosophy of Martin Luther King Jr.* There they write that "despite King's having been memorialized so widely and quoted so frequently, serious study and criticism of his writings, speeches, and sermons remain remarkably marginal and underdeveloped within philosophy, political theory, and the history of political thought—even in those subfields where one might expect his contributions to be essential reading."[7] They also argue that the "ritual celebration and intellectual marginalization" of King is connected to a "romantic" "master narrative" of the Civil Rights Movement that is limited thematically, temporally, and geographically, as well as in terms of its heroic cast of characters, which focuses too much on ideas of "national unity" and "transcendence over the evils of racial oppression."[8] King's thought can at times be interpreted as a contradictory narrative in which Black Power is the result of the multivalent failures of the Civil Rights Movement, and, alternatively, as one of (potential) romantic closure, in part because of the influence of his oft repeated statement "that the arc of the moral universe, although long, is bending toward justice."[9] However, King explicitly situates the civil and human rights movements in the United States in relation to the global decolonization movements, and his foregrounding of right-wing political reaction troubles any easy understanding of King's sense of postracial futurity. Most important, an examination of King's thought demonstrates the extent to which conceptualizing the politics of automation and Black Power were mutually constitutive endeavors. It also further broadens our understanding of the ways in which King, and others in the period, theorized cyberculture and technological change in terms of the continuing influence of the plantation economy and the distributions of waged and unwaged work, and freedom and unfreedom, that were both new and historic and that demanded an anticapitalist reorganization of production and social reproduction in the face of intractable forms of political-economic inequality.

King's engagement with automation and technology was a concern that dates to the earliest moments of his public political life.[10]

This is evidenced, for example, in sermons like "Paul's Letter to American Christians," delivered on the eve of the conclusion of the bus boycott at Dexter Avenue Baptist Church in Montgomery, Alabama, on November 12, 1956, and "Facing the Challenge of a New Age" at the NAACP's Emancipation Day Rally in Atlanta on January 1, 1957. The latter uses many of the same rhetorical themes as the 1968 "Remaining Awake Through a Great Revolution" sermon. The systematic, while also recursive and nonlinear, development of King's engagement with the politics of automation is especially brought out through an examination of what I will call his "Remaining Awake Through a Great Revolution" works, which include not only "Facing the Challenge of a New Age" but also his final book, *Where Do We Go from Here: Chaos or Community?*.

Throughout these, we find King working through a well-defined set of rhetorical phrases and political concerns. However, these were continually reworked as political conjunctures changed. A closer comparison of the 1959 Morehouse commencement address version of "Remaining Awake Through a Great Revolution," the first time he used that title, and the 1968 version will illuminate some of their parallels and disjunctures while also demonstrating the extent to which technology and political economy were central to his thought.

Both speeches start with an anecdote King frequently used about the Washington Irving character of Rip Van Winkle. In the June 1959 Morehouse speech, he begins:

> There can be no gainsaying of the fact that we are experiencing today one of the greatest revolutions that the world has ever known. Indeed there have been other revolutions, but they have been local and isolated. The distinctive feature of the present revolution is that it is worldwide. It is shaking the foundations of the east and the west. It has engulfed every continent of the world. You can hear its deep rumblings from the lowest village street to the highest intellectual ivory tower. Every segment of society is being swept into its mainstream. The great challenge facing every member of this graduating class is to remain awake, alert and creative through this great revolution.[11]

He begins the 1968 version in a similar way. Both make the argument that the main point was "not that he [Rip Van Winkle] slept twenty years, but that he slept through a great revolution." The revolutions that King refers to in the 1959 speech are specifically the Bandung-era anticolonial ones in Asia and Africa. The challenge that King makes to the graduates of Morehouse is that they must not only be attentive to how these revolutions were connected to the emerging Civil Rights Movement but that they must also be "creative" in their response to both. There are other rhetorical and narrative parallels between this version and the 1968 iteration, like the story of his trip to India in March 1959. However, there is a significant difference in the way that the texts approach automation and technology change. I quote at length from the Morehouse commencement address because the language is so evocative of the space age:

> There is not only a revolution taking place in the social and political structure of man's being, but there is a revolution taking place in the external physical structure of his being. In other words, a revolution is taking place in man's scientific and technological development. Man through his scientific genius has been able to dwarf distance and place time in chains. He has been able to carve highways through the stratosphere, and is now making preparations for a trip to the moon. These revolutionary changes have brought us into a space age. The world is now geographically one. Jet planes have compressed into minutes distances that a few years ago took weeks. Bob Hope has described this new jet age in which we live. He says it is an age in which we will be able to take a non stop flight from Los Angeles to New york city, and if by chance we develop hiccups on taking off, we will "hic" in Los Angeles and "cup" in New york city. It is an age in which one will be able to leave ~~tokyo~~ Tokyo on Sunday morning and, because of the time difference, arrive in Seattle Washington on the preceding Saturday night. When your friends meet you at the airport in Seattle inquiring when you left Toyko you will have to say, "I left tomorrow." This is a bit humorous, but it reminds us that a great revolution is taking place in the physical structure of our universe.
>
> Now the great question facing us today is whether we will remain awake through this worldshaking revolution, and achieve the new mental attitudes which the situations and conditions demand.[12]

Here King emphasizes, through the language of Bob Hope, what David Harvey calls the "time-space compression" of technocapitalist modernity.[13] Technology is secondary, at least in its placement, in the 1959 speech, but in 1968 it moves to the forefront, where it is no longer framed through the language of "Hope" and the possibility that the technological domination of nature can produce human freedom. Instead, it is conceptualized through the language of the triple revolution, foregrounding the importance of automation to the politics of equality and poverty in 1968, as well as an increased sense that technological change and the possibility of freedom is dialectically linked to rightist counterrevolution and the recurrence of slavery's histories:

> And now if we are to do it we must honestly admit certain things and get rid of certain myths that have constantly been disseminated all over our nation. One is the myth of time. It is the notion that only time can solve the problem of racial injustice. . . . There is an answer to that myth. It is that time is neutral. It can be used wither constructively or destructively. And I am sorry to say this morning that I am absolutely convinced that the forces of ill will in our nation, the extreme rightists of our nation—the people on the wrong side—have used time much more effectively than the forces of goodwill. And it may well be that we will have to repent in this generation.[14]

In a broad sense, this chapter asks what is at stake for our understanding of Black Power cyberculture in reading these two different "Remaining Awake Through a Great Revolution" speeches. On the one hand, they are largely consistent in their language and in their concerns with the period's political, technological, and economic revolutions. Perhaps unsurprisingly, the 1968 speech is suffused with the urgency of rightist counterrevolution, the breaking down of Keynesian capitalism's ability to reproduce itself, and, although not explicitly stated, the presence of urban uprisings and the Black Power movement. If the *now* can no longer be mediated by Bob Hope, and if the immediacy of the present is that much more urgent, then it is also a speech that, as I will elaborate below, is about how the United States "undergirded" White Power and capitalism through its support for technologies of exploitation and dispossession, like farm mechanization. Above all, what this as a first, brief example demonstrates is that to read across these speeches is to see the ways that the legitimation crises of the cybercultural era had affected King's thought. Clearly the legitimacy of the U.S. racial regime was being called into question by the Montgomery bus boycott and the emergence of the Civil Rights Movement in 1959. However, by 1968, the renewed fervor of the U.S. racial regime, combined with the undermining of Keynesian forms of production and social reproduction by both real and perceived forms of automation, were intensifying the crisis. Charting King's thought helps us see how the problem of automation was not simply a crisis in White male unemployment but was also part of a much broader struggle over the social reproduction of capitalism and its alternatives, which called into question the goals of a civil rights movement that was increasingly out of sync with emergent forms of unemployment, inflation, geographic reorganization, and antiwelfare politics, all of which made integration into a still racially unequal capitalism that much more unlikely. In the face of these, King would call for a universal guaranteed income and an interracial Poor People's Campaign that would rethink gender and race, wage labor, the work ethic, human value, how capitalism reproduced inequality, and the types of massive political movements necessary to reorganize society. In order to set the stage for this, I turn first to an examination of the limits and possibilities of Keynesian state capitalism in the face of automation and cybernation by analyzing *The Triple Revolution.*

The Triple Revolution and the New Science of Political Economy

Examples of the period's cybercultural productions range from episodes of *The Twilight Zone* and *Star Trek* to the Students for a Democratic Society's *Port Huron Statement* and Shulamith Firestone's *The Dialectic of Sex.* Two of the most influential documents were *The Triple Revolution* by the Ad Hoc Committee on the Triple Revolution, which was published as an open letter to President Lyndon Johnson in March 1964, and the National Commission on Technology, Automation, and Economic Progress (NCTAEP), created by the U.S. Congress in April 1964 and signed into law by Johnson in August. They published their first report, *Technology and the American Economy,* in February 1966.

Of the two, *The Triple Revolution* received the most sensational reviews. For example, an August 1964 *Life* magazine editorial, "If the Machine Wants Our Jobs, Let's Buy It," referred to its authors as a "group of far-out thinkers." In a review, replete with space-age clichés, critical of what it saw as the committee's "radical" bent, *Life* called for a solution to automated unemployment, one that recognized the "success" of capitalism in creating automation technologies and that would parallel the land distribution policies of the Homestead Act, perhaps surprisingly "plac[ing] ownership of U.S. industry in the hands of the workers and let[ting] them live off their dividends as their job incomes fall."[15] Martin Luther King Jr. saw a different relationship than *Life* between the Homestead Act and technological development because, as he points out, technology change in U.S. capitalism up to that point had always been premised on dispossession and the production of inequality and White Power. However, *Life*'s worker-owned response to automated unemployment speaks to the historical conjuncture that I am constructing here as one where the assumptions of Keynesian capitalism were being broadly called into question by automation, even in publications like *Life*.

Because *The Triple Revolution* was used by King as a framing device to think through the ways that racialized difference, especially, was constitutive of the political histories and horizons of the cybercultural era, and because James Boggs was one of its dissenting authors, I primarily focus on it here. However, taken together, the AHC and the NCTAEP offer a clear sense of the Keynesian politics of a potentially "cybernated" economy—*cybernation* defined here as automation and/or mechanization with computers. Most significantly, the two saw the immediate features of automation differently. The AHC argued that with the "Cybernation Revolution," "a new era of production has begun. Its principles of organization are as different from those of the industrial era as those of the industrial era were different from the agricultural. The cybernation revolution has been brought about by the combination of the computer and the automated self-regulating machine. This results in a system of almost unlimited productive capacity which requires progressively less human labor. Cybernation is already reorganizing the economic and social system to meet its own needs."[16] According to the AHC, this emergent radical reorganizing of society would require that a "new science of political economy will be built on the encouragement and planned expansion of cybernation."[17] Alternatively, the NCTAEP stated, "Our broad conclusion is that the pace of technological change has increased in recent decades and may increase in the future, but a sharp break in the continuity of technical progress has not occurred, nor is it likely to occur in the next decade."[18] Despite the apparently dramatic difference between the two over whether or not automation marked an actual break with past modes of production, the two agreed on a great deal, including a recognition that technological change, like farm mechanization and factory automation, had resulted in many forms of displacement, and that many of those impacts were racialized and unequal. They also shared a horizon of the future in which technology could potentially offer hitherto untold benefits to humanity, if managed properly, and they agreed that people should be provided with a guaranteed income in recognition of expected disruptions in the job market—an idea that the two documents helped to make popular enough that Richard Nixon incorporated it into his 1968 presidential campaign platform.[19]

This was one of the features of what Howard Brick describes as the postcapitalist consensus of the 1950s and 1960s in the United States. In an era of technological change, affluence, and abundance, in which leisure and service industries were on the rise and industrial labor was on the decline, there was a broad agreement about the need to collectively manage social and economic affairs.[20] Melinda Cooper has demonstrated that this sense, which she describes as one of "cautious technocratic optimism," included protoneoconservative journals like *The Public Interest* and *Commentary* that broadly accepted the New Deal and Great Society welfare state, including the potential inclusion of Black men–led nuclear families into its fold.[21] The AHC and the NCTAEP testify to the mid-1960s consensus in favor of managing technological change in a postindustrial consumer society. This was especially necessary because both documents evidence a shared concern that automated technological change was altering the foundations of the post–World War II Keynesian organization of effective demand, full employment, the family wage, and the production of Americans, across racial categories, as consumers. However, they did not share assumptions about the immediacy of change or the extent of the break with the past that automation represented, and thus whether it needed to be confronted with a sense of urgency, for *The Triple Revolution,* or with a moderate technocratic approach, in the case of the NCTAEP. In this respect, *automation* in the mid-1960s signaled the struggles that were to soon come over social reproduction and whether the social safety net should be extended beyond the normative White nuclear family.

Although automation, or technological unemployment, has been a recurrent problem in the history of capitalism, as a term, *automation* was in fact relatively new in the 1960s. In 1956, Friedrich Pollock began *Automation: A Study of Its Economic and Social Consequences* with, "'Automation' is a new word in the English language" with a "number of different meanings, ranging from conveyor-belt production to highly complicated forms of automatic machinery" and with synonyms like "cybernetics" and "automisation." Most significant, according to Pollock, it was "now ousting the other words as an expression denoting a technical development that is replacing human labour by machinery in factories and workshops in a way that would have been thought impossible only ten years ago."[22] *Automation* was coined in 1946 by Ford's Del Harder, who was hired to organize its River Rouge complex, which would be the location of much LRBW activism, and set up its automation department. As Jason Smith notes, by *automation,* Harder was not referring to Norbert Wiener's cybernetic ideas or to computer-controlled machines. Instead, "he meant merely the growing preponderance of what he called 'electro-mechanical, pneumatic, and hydraulic' devices in factory production," most of which dated to the nineteenth century.[23] The uncertainty and newness surrounding the term is also evidenced by Howard Bowen, head of the NCTAEP, who wrote in the popularization of the commission's report, *Automation and Economic Progress,* that "'automation' is a catchy term" and that "in common parlance it became synonymous with any force having a negative effect upon employment."[24] Although it might be obvious, *automation* as a concept thus marks the post–World War II emergence of the political problem of technological unemployment, especially in the United States. Aaron Benanav argues that there are four features of what he calls the automation discourse. First, there is already "technological unemployment" due to increasingly advanced machines. Second, this is a sign that society is on the verge of becoming a largely automated society in which all work will be performed by self-moving machines and intelligent computers. Third, this will most likely turn out to be "a nightmare," even though in principle this could mean humanity's liberation from a society in which most people have to work in order to live. Finally, the only way to prevent a mass-unemployment catastrophe is to institute a universal basic income (UBI) that will break the link between work and income.[25] All of these features are present in *The Triple Revolution;* however, the political problem of unemployment, while a general feature of capitalism, was also specific to the United States in the 1960s in two ways. First, high unemployment rates in the 1930s, labor struggles, and the rise of fascism weighed heavily on the era's response to actual and anticipated automated unemployment. Second, the civil rights struggle sought to incorporate, especially, Black men into the welfare state and a racially segregated Fordist labor force during a time of high unionization rates and correspondingly high wages.

Perhaps coincidently, 1946, when Harder first used the term *automation,* was also the year the U.S. Congress passed the Employment Act of 1946. Initially called the Full Employment Bill, its ultimately stated aim was to "promote maximum employment." *The Triple Revolution* was written in response to the limits of the War on Poverty and the soon-to-be-signed Economic Opportunity Act of 1964 and was primarily concerned with full employment and the corresponding problem of effective demand. It believed these were compromised by cybernation as well as racial segregation, and it called the specter of unemployment in the United States the "stage on which the machines-and-man drama will first be played for the world to witness."[26] In fact, even the more conservative NCTAEP recommended that the government "fulfill the promise of the Employment Act of 1946" and make "the Government be an employer of last resort," and it argued for the "elimination of all social barriers to employment and advocating special programs to compensate for centuries of systematic denial," in addition to other recommendations.[27] While primarily a document of technoliberalism focused on the risks posed by automation to a universal human, in part because *The Triple Revolution* is coauthored by James Boggs, Bayard Rustin, and others, but especially because of the Black freedom movement, the unequally racialized impacts of automation on employment are not entirely absent, even if it was seen as a secondary effect and not constitutive of technological development itself.

Although the years in which *The Triple Revolution* and the NCTAEP's reports were issued were ones during which official unemployment was decreasing, the AHC argued that "Unemployment is far Worse than Figures indicate" and that the "private sector" was actually creating virtually no new nonservice jobs in a U.S. economy that overly relied on military spending to support the system as a whole. According to the AHC, the low official unemployment rate hid the fact that there was emerging a "permanently depressed class" made up of the old and the young, as well as Black Americans, who, through multiple forms of segregation, were denied access to employment. This unemployment problem, which might normally be addressed in various ways, such as through social spending increases or tax cuts to stimulate hiring (which was done in the 1964 Kennedy–Johnson tax cut), was being radically transformed by the cybernation revolution that

> invalidates the general mechanism so far employed to undergird people's rights as consumers. Up to this time economic resources have been distributed on the basis of contributions to production, with machines and men competing for employment on somewhat equal terms. . . . As machines take over production from men, they absorb an increasing proportion of resources while the men who are displaced become dependent on minimal and unrelated government measures—unemployment insurance, social security, welfare payments. These measures are less and less able to disguise a historic paradox: That a substantial proportion of the population is subsisting on minimal incomes, often below the poverty line, at a time when sufficient productive potential is available to supply the needs of everyone in the U.S.[28]

They believed that there was a crisis in male employment as well as in the mechanisms through which the Keynesian "industrial" system managed the contradictions of capitalism in order to secure (near) full employment and effective demand. As they further claimed, there is

> no question that cybernation would make possible the abolition of poverty at home and abroad. But the industrial system does

> not possess any adequate mechanisms to permit these potentials to become realities. The industrial system was designed to produce an ever-increasing quantity of goods as efficiently as possible, and it was assumed that the distribution of the power to purchase these goods would occur almost automatically. The continuance of the income-through jobs link as the only major mechanism for distributing effective demand—for granting the right to consume—now acts as the main brake on the almost unlimited capacity of a cybernated productive system.[29]

One of the most common narratives of neoliberalism argues that it emerged in the mid-1970s as a response to contradictions in Keynesian capitalism like stagflation and overaccumulation. *The Triple Revolution* suggests a shadow history to that emergence—not a contradictory path but a historically alternative one, which parallels the work of Melinda Cooper and David Harvey.[30] As Harvey writes in *The Condition of Postmodernity,* "Postwar Fordism . . . [was] less a mere system of mass production and more a total way of life."[31] This included the standardization of mass consumption and the preservation of the work ethic. Cooper adds to this that the normative family model is essential to this because of the ways it was organized to regulate desire, in multiple senses.[32] As they both argue, for people like Daniel Bell, the main risk of the welfare system is that it would free people from the constraints of the work ethic and the normative family—if, that is, it were to be expanded beyond those normative constraints. This risk was present because the many different countercultures of the 1960s—Black women, single mothers, gay liberation—sought access to benefits largely available only to households headed by White men. Jason Smith and Aaron Benanav argue that the rhetoric of automation in the 1960s, as well as today, often overstates the actual links between automation and unemployment. However, most important, *The Triple Revolution* points to the political problem of automation in the 1960s. First, it recognized that mass automated unemployment would immediately require new ways of stimulating effective demand beyond the family wage. Second, it suggested that something more liberating might develop out of cybernation if it could be directed properly—that is, when labor and life are disarticulated, which is the precise fear, according to Cooper, of "the new social conservatism."

Thus, while framing itself as a problem of "machines tak[ing] over production from men," *The Triple Revolution* in fact opens the problem of social reproduction in a midcentury U.S. capitalism that had relied on (near) full White male employment and on a hierarchically racialized family to reproduce the labor power and rationalized forms of consumption that industrial capitalism required, along with its correlate: a largely racialized and minoritized surplus labor population. This is important because the cybercultural era is best defined not as a technologically determinist moment in the linear unfolding of automation's development but rather as a conjuncture in which (potential) automated unemployment was closely and necessarily connected to the racialized and gendered politics of the wage, welfare, and the reproduction of capitalism. *The Triple Revolution,* and in a different way the official NCTAEP, are representative of this as they bring into relief the possibilities and limits of the U.S. Keynesian welfare-warfare state at the moment of the War on Poverty, the Great Society, the Cold War, the war in Vietnam, and multiple freedom struggles. Black Power cyberculture, in different ways, formulated its politics against not the most limited responses by technoliberalism to unemployment and racialized and gendered inequality, but against their most expansive formulations, like *The Triple Revolution.*

The Triple Revolution's response to the (coming) crisis of Keynesian reproduction was not the market-based, entrepreneurial approach called for by the new social conservatism. Instead, it argued that a planned "Transition" involving "massive" programs including public housing and public works (like public coal-powered electricity), increases in educational funding and public infrastructure, revision of the tax structure, unionization of the unemployed, and government regulation of the speed and implementation of cybernation, among other programs, was necessary.[33] Their most far-reaching proposal, paralleling Benanav's fourth feature of the automation discourse, was to be "an unqualified commitment to provide every individual and every family with an adequate income as a matter of right." In a world in which "the traditional link between jobs and incomes is being broken" and "wealth [is] produced by machines rather than by men," such a program was both necessary and just.[34] As I expand on below, a guaranteed income offered the possibility of breaking the links between waged labor and human value demanded by a capitalist world of (organized) scarcity. At the same time, it was, and remains, a highly political concept that can be used in the opposite way too. For example, Milton Friedman called for a negative income tax in the 1960s to provide an income threshold for all, but it would also be accompanied by the destruction of other social welfare programs while also being used as a tool to more fully integrate people into the "free" market—a sense of the UBI shared by many in Silicon Valley today.

Benanav argues that the automation discourse arises in moments "when the gap between the supply and demand for jobs becomes so large . . . that people begin to question the viability of a market-regulated society." This "anxiety" can be productive, opening a space to think through the "limits" of capitalism and at their best these discourses can point to, and struggle for, the "utopian possibilities latent within capitalist societies."[35] For *The Triple Revolution,* "establishment of the right to an income," unlike other proposals in the period, did not call for the elimination of other forms of welfare and for the incorporation of worker-entrepreneurs-of-the-self more efficiently into market, but "will prove to have been only the first step in the reconstruction of the value system of our society brought on by the triple revolution."[36] Here automation is the limit case for imagining a new human and society:

> In the absence of real understanding of any of these phenomena, especially of technology, we may be allowing an efficient and dehumanized community to emerge by default. Gaining control of our future requires the conscious formation of the society we wish to have. Cybernation at last forces us to answer the historic questions: What is man's role when he is not dependent upon his own activities for the material basis of his life? What should be the basis for distributing individual access to national resources? Are there other proper claims on goods and services besides a job?[37]

At the same time, this new human and society would require a new science of political economy that would be both managed and also democratic, consisting of people who are able to "understand, express and determine their lives as dignified human beings" because wealth is distributed with "the widest possible social benefit."[38] This, however, is a thirteen-page political pamphlet that is neither about the political practices that would create this world nor about the technologies of capitalism that extract surplus value and maintain political power through the exploitation and production of social difference.

These absences were central to *Monthly Review*'s critique of *The Triple Revolution*.[39] An editorial review argued that, despite the document's strengths, it was capitalism as a system, and not technology, that caused unemployment and prevented people from having more leisure time. Further, a universal income, expanded leisure rights, and a postscarcity world could be had without the precondition that automation made it necessary. Politicizing the assumptions of automation and a postlabor, postscarcity world raised the question of the transition to better (or worse) political economies—something that Martin Luther King Jr. was acutely aware of in 1968. In *One-Dimensional Man*, Herbert Marcuse referred to the welfare-warfare state as a "historical freak" trapped, purposefully, between "capitalism and freedom," semiautomation and full automation, and abolishing labor and preserving it, as well as between "servitude and freedom" and "totalitarianism and happiness."[40] *The Triple Revolution* shares something of this in that 1964 is like an interregnum between the old social system and the new. However, it is explicitly future oriented in its horizon—a future of possibility but also one of crisis in a contradictory technocapitalism unable to properly reproduce itself while it pushes beyond its internal limits. Because of this, like much of today's accelerationist thought, the AHC suggests it is only by embracing and traversing a (managed) cybercultural revolution that a possible future can arise.[41]

How, though, in the technoliberal world of *The Triple Revolution* is this revolution to be politically made? This is a problem that haunts the AHC and that is brought into relief in a talk called "Education and the Triple Revolution" by W. H. Ferry, vice president of the Center for the Study of Democratic Institutions. In language at times echoing Reinhart Koselleck's characterization of modernity as one in which there is a gap between the "space of experience" and the "horizon of the future," he described his present as one of "relative velocities."[42] He further noted:

> The task of twentieth century education is to bring social and political imagination into workable parity with scientific and

> technological imagination. I am not at all certain that this is feasible. The evidence so far is against it. Accelerating change seems always to mean accelerating crisis. We are running an outer space civilization on a farm-based Constitution and with national ideology more appropriate to nineteenth century Europe than to twentieth century America. Politics is far out of step with technology. The gulf widens daily.[43]

This passage effectively condenses *The Triple Revolution*'s modernist sense of accelerated change and the cybernetic untethering of the material constitution of society, its forms of production and social reproduction, from its formal constitution. James Boggs dissented from the recommendations of *The Triple Revolution* precisely because he believe that they were too limited: they were not geared toward the reality of cybercultural revolution and its production of outsiders; nor were its politics equal to the task of making the new world, and the new human, it foresaw. Initially Martin Luther King Jr., perhaps necessarily, formulated his response to the politics of automation primarily in a Keynesian family-wage frame, modified by his critique of its racial underpinnings. This would begin to shift through dialogue with the Black Power movement, especially as it was represented in urban rebellions, and dialogue with the Black women activists of the National Welfare Rights Organization and their expansive sense of a guaranteed income. Ultimately, by 1968, when he framed his politics in terms of *The Triple Revolution*'s problematic, he would argue that a true postscarcity society of equality, made only in part possible by automation technologies, required mass political mobilization and new solidarities, as well as a reckoning with the histories of capitalism, its necessary inequalities, and its continual racialized and gendered differentiation between waged and unwaged work and freedom and unfreedom.

The Plantation Machine, Automation, and the General Economy of the Slum

In March 1966, at the Chicago Freedom Festival, Martin Luther King Jr. described the Black urban condition as the "General Economy of the slum": a world continually "drained," "exploited economically," and "dominated politically" in which Black homeowners and renters paid substantially more for lower-quality homes purchased at higher interest rates.[44] This was taking place in the midst of a "depression" in which the official Black unemployment rate had risen from 4.9 percent in 1953 to 10.9 percent in 1963. King argued that Black, primarily male workers displaced from agriculture in the South and seeking freedom in the North were now once again the first "displaced by automation and plant move-outs" and were inevitably becoming a permanent underclass of the unemployed. The general economy of the slum is another way to describe the negative sense of the cybercultural revolution. Through it, King analyzed a specific transitional moment in the history of racialized value extraction in the United States that was not only about dispossession in Black urban neighborhoods but about the ways that capitalism was reorganizing itself geographically and new forms of racialization and class composition were emerging. He connected automated dispossession to the historic forms of value derived from low-waged and unwaged work, like agricultural and manual labor in the "plantation machine" and its long afterlives, to what Keeanga-Yamahtta Taylor has described as a new condition of "predatory inclusion" where the development of subprime, Federal Housing Authority–backed loans created a market in which Black dwellings were organized as exchange value in a dialectical relationship to White suburban communities in which dwellings were "homes" with use value.[45] Alongside high unemployment rates and high-interest loans for substandard dwellings, Black communities like the one King lived in in Chicago were also subject to extraction and precarity through multiple race taxes. In sum, Chicago, and Black neighborhoods in other Northern cities, were, King said, "little more than domestic colonies" (drawing on one of the Black Power movement's key political economic discourses).

The reorganization of capital and the emergence of Black Power also required King to reconsider the modalities, and the subjects, of the political struggle for racial equality and a just distribution of society's surplus. While supporting striking sanitation workers in Memphis in 1968, King elaborated on the political urgency of struggle against these forms of dispossession, stating before the American Federation of State, County and Municipal Employees that the movement was "going beyond purely civil rights to questions of human rights."[46] This was not just another way of rephrasing legal language. Instead, it was a necessary shift to the material constitution of society because by focusing on human rights one was "highlighting the economic issues. . . . That is a distinction."[47] Human equality and economic equality were thus synonymous. This shift is what King called the second phase of the struggle—one in which the stakes were higher because White allies were disappearing and a formidable White counterrevolution was making itself present because real costs, both financial and psychological, would now have to be borne by U.S. society in order to create a world of equality.

Reading across Martin Luther King Jr.'s speeches and writing from the late 1950s to his death reveals that this analysis is representative of an extended engagement with the relationship between the politics of technology change and the production of racial difference and inequality in U.S. and planetary capitalism. King's political economic thought was grounded in the pressing contingencies of the present and how they were connected to, and were necessarily rethought, in terms of capitalism's longer durée. King shares *The Triple Revolution*'s call for extensive planning and a reconceptualization of what it meant to be human when value is dearticulated from wage labor, and, for King, the histories of racialized value, at a moment in which automation technologies appeared to be reorganizing society. However, King insists in his Chicago Freedom Festival speech, as he did before labor unions like the American Federation of State, County and Municipal Employees, that automation was a political problem that, although in many respects new, was continuous with the afterlife of slavery. While automation exacerbated Black unemployment and inequality in Northern cities, automation and inequality were also constitutive of each other, part of capitalism's necessary logic of exploiting forms of social difference and producing a largely racialized reserve army of labor. The demand to rethink human value for an automated world required an examination of what was not only novel about the potentials of automation but also the extent to which automation, as a key term in the history of capitalism's general economy of the slum, takes us to the constantly rearticulating boundaries between waged work and wagelessness, freedom and unfreedom, the human and the less than fully human, and inequality generally on which White Power and capitalist political economies are based, and which racial capitalist and feminist critiques have been especially productive at explicating. Charting King's thought from the late 1950s until his engagement in the late 1960s with Black Power and domestic colony theory is especially useful because of the way it mediates debates about the organization of U.S. and planetary capitalism as part of a political movement that sought full access to the affluent society, but that increasingly began to argue that the combined history of technocapitalist inequality, new forms of technological unemployment, and White rightist reaction put that access out of reach and demanded a more radical reimagining of human value beyond the limits of Keynesian affluence.

The intersection of racialized economic inequality and the politics of technological change were increasingly urgent problems for King in the late 1960s, but they had been crucial features of his thought since the 1950s. Publicly naming technocapitalism as a system of racialized inequality dates at least to his sermon "Paul's Letter to American Christians" delivered at Dexter Avenue Baptist Church on November 4, 1956, just before the end of the Montgomery bus boycott on December 21.[48] Here, echoing the voice of Paul in his letters to the Romans, he talks, as in the speeches that introduced this chapter, about the "tremendous strides" in "scientific and technological development" that allowed speedy travel over great distances and that had created an interconnected world—indeed, these passages reverberate with the bus boycott's political struggle over full access to transportation technologies. Yet as in those other speeches, he questions whether America's "spiritual progress has been commensurate with your scientific progress." He attributes this gap to the "economic system in America known as Capitalism," the "misuse" of which leads to "tragic exploitation" where "one tenth of one percent of the population controls more than forty percent of the wealth" and where success is judged "by the index of your salary and the size of the wheel base on your automobile, rather than the quality of your service to humanity"—though he does say that "democracy" still has the potential to modify for the better capitalism's inequalities. This speech shows that in King's thought, the politics of technology were always intertwined with capitalist inequality, even if here the full implications of technocapitalism are not fully elaborated. This is true as well in King's first use of the term *automation,* which occurred in February 1958 in his "Remarks for Negro Press Week."[49] There he said that for "future historians" looking back on the twentieth century "long after man has adjusted to the atomic age and even after the problems of automation have been dealt with, our age will be remembered as the period when ordinary men demanded to live with dignity and freedom." Here automation seems to be a problem at least as grave as that of nuclear weapons; however, how this is so is left to the imagination.

In between these first two public statements on capitalism and automation, King delivered a speech at the Highlander Folk School in which he outlined a three-phase history of U.S. political economy. Here he begins to make explicit the connection between plantation slavery and technology as a racialized form of extraction and devaluation, which would be increasingly central to his political economic thought leading up to the Poor People's Campaign. The three phases of history are not unique to King: the first phase of 1619–1863 was that of plantation slavery, the second of 1863–1954 of segregation, and the third of the then present "period of complete and constructive integration."[50] Significantly, he describes the period of plantation slavery as one where Black Americans were "an 'it' rather than a 'he,' a thing to be used rather than a person to be respected . . . merely a depersonalized cog in a vast plantation machine"[51]—a formulation that he also used in one of the first iterations of his "Remaining Awake" speeches, also from 1957, at the NAACP Emancipation Day Rally entitled "Facing the Challenge of a New Age."[52] King's description of the plantation as a depersonalizing "machine" that produced "its" and "things" will reverberate in his repeated calls for the economy to be transformed from one that is "thing-oriented" to one that was "people-oriented."[53]

This resonates with what Louis Chude-Sokei calls a "creole" history of race and technology. Chude-Sokei argues in *The Sound of Culture: Diaspora and Black Technopoetics* that in modernity, *race* and *technology* have always been mediating and contingent terms, whether in science fiction or Black music, and this was especially true of the plantation. The metaphor of the "plantation machine" was used by the planter class in order to highlight the efficiency and rationality—which is to say the modernity—of the plantation. For example, Samuel Martin of Antigua wrote in the 1765 *An Essay upon Plantership* that "a plantation ought to be considered as a well-constructed machine."[54] It was also taken up by Black diasporic and anticolonial critical thought about the plantation by C. L. R. James, Sylvia Wynter, Aimé Césaire, and Antonio Bénitez-Rojo, among others. For C. L. R. James, the plantation was the forerunner of the modern factory because its forms of "regimentation" had made enslaved peoples a "modern proletariat" capable of revolutionary action, while for Sylvia Wynter the plantation was a machine that colonized "desire."[55] Especially close to King's own conceptualization of the plantation as a thing producing machine is Césaire's famous formulation in *Discourse on Colonialism* that "colonization = thingification." For Césaire, this is because the discourse of civilization is undergirded by practices of colonization that included "forced labor," "rape," "compulsory crops," "submission," and, perhaps most important, of "extraordinary possibilities wiped out," through which people were transformed into less than fully human, colonial subjects.[56] King's formulation parallels Césaire, Wynter, and James in examining the ways in which capitalism is a racial formation that thingifies people through the exploitation and production of forms of hierarchical difference in order to extract surplus value while simultaneously wiping out other potential ways of being in the world than unequal, capitalist ones. In this respect, King's political economic thought asks similar questions to those that in part animate Jennifer Morgan in *Reckoning with Slavery,* where she examines how difference, value, and economic rationality emerged as interrelated issues with Atlantic slavery. King saw the history of automation in the 1960s and its new forms of dehumanization and value extraction as necessarily linked to the history of the plantation machine's production of value as solely a question of accounting—that is, of money and capital.

King's political response to thingification and racialization is almost always described as integrationist, and this is undoubtedly true, as his periodization of the historical present suggests. Detractors might argue that naive mainstream claims about integration (which is not to say that King's politics of integration was naive) paralleled technoliberalism's claims about a universal human for an age of technocratic American affluence. Yet although integration was always an aspiration, by "Remaining Awake" and *Where Do We Go from Here?,* that long arc of history seemed to no longer hold. Instead, King argued that time was beginning to move backward and, as I quoted above, that there was a counterrevolution of "extreme rightists . . . [who] have used time much more effectively than the forces of goodwill." Counter to mainstream cyberculture, the automation revolution, instead of potentially creating a new America of freedom and affluence beyond labor, seemed to be producing new forms of human obsolescence that extended the afterlives of the plantation machine's thing-producing technologies. Automation for King thus demanded a rethinking of the histories of capitalism and the trajectories of the Black freedom struggle. As he wrote in *Where Do We Go from Here?:*

> When the Constitution was written, a strange formula to determine taxes and representation declared that the Negro was 60 percent of a person. Today [1967] another curious formula seems to declare he is 50 percent of a person. Of the good things in life he has approximately one-half those of whites; of the bad he has twice those of whites. Thus, half of all Negroes live in substandard housing, and Negroes have half the income of whites. When we turn to the negative experiences of life, the Negro has a double share. There are twice as many unemployed.[57]

Saidiya Hartman has described the "time of slavery" as one that "trouble(s) redemptive narratives" about the pastness of slavery and that "negate the common-sense intuition of time as continuity or progression."[58] By reconsidering the development of capitalist inequality in the late 1960s, King called into question any straightforward, linear trajectory of freedom. For King, the moment of the plantation machine is one of primitive accumulation; it is one in which enslaved African peoples were violently uprooted from their lifeworlds and turned into commodified, and racialized, productive and reproductive things. The cybercultural era of counterrevolution and continued impoverishment, in which the pain of capitalism was unevenly distributed in a situation of spectacular deprivation relative to an affluent society, thus signifies the continued presence of forms of primitive accumulation, racialized exploitation, and surplus-value extraction.[59]

Reading Marx alongside developments in racial capitalist theory, Nikhil Pal Singh argues against versions of capitalist history that claim slavery and primitive accumulation are past or antiquated forms that cease to exist, or at least be relevant, when wage labor and market-based commodification become the so-called norm. Instead, he argues that the interrelationship between slavery and freedom and waged and unwaged labor is an ongoing constitutive feature of capitalism: "Capital ceases to be capital without the ongoing differentiation of free labor and slavery, waged labor and unpaid labor. This differentiation provides the indispensable material and ideological support for capitalism's continued development. The absolute separation of freedom and slavery operates in the interests of capital. It is only by retaining an understanding of their overlapping dimensions that we attain a critical perspective adequate to oppose it."[60] The 1968 version of "Remaining Awake Through a Great Revolution," like in his Chicago Freedom Festival speech, critiques what he called the "bootstrap philosophy" of freedom espoused by, especially, White liberals, where freedom is described as developing naturally over time through individual struggle, self-realization, and wage labor in a world free of White Power. The intersection of the automation and human rights revolutions are thus embedded in a general economy of racial capitalism where freedom through wage labor is not fully realized. Automation as a machine of unemployment—one that troubles the boundaries of freedom and unfreedom as well as waged and unwaged work—thus emerges out of the plantation machine's production of racialized inequality.

For King, the political history of automation was not simply technological in the sense of the machine but also technological in the sense of multiple political techniques that organized life as racialized and thingified during the plantation period and that continued to undergird White Power in the post-Reconstruction era. Again in "Remaining Awake Through a Great Revolution," he attributes the impoverishment exacerbated by the automation and cybernation revolution, unlike *Life* magazine, to the history of the Homestead Act, which dispossessed Indigenous peoples at a time in which the formerly enslaved were being denied land and full legal equality. The period of the Homestead Act also saw the creation of land-grant universities, which in his specific analysis were schools for training emergent industrial farmers, whose farming techniques, alongside commodity crop subsidies and funding for farm mechanization technologies, combined to displace Black workers from the land. All of these consolidated Whiteness itself by providing people, especially newly arrived European immigrants, with an "economic floor" at a time of continued primitive accumulation by White capitalism; indeed, King's use of the phrase "economic floor" mirrors that of the period's advocates of a negative income tax.

King thus mobilized a longer durée in order to contextualize the immediate racial and class crises of the triple revolution. Alongside this, the triple revolution and the general economy of the slum were also complex geographic problems. This is explicit in his "Remaining Awake" sermons, as well as the speech he delivered at the Chicago Freedom Festival. It was also a consistent theme in speeches he gave to labor unions. In them, he attempted to move major unions like the United Auto Workers from a place of public, but limited, support for the Black freedom movement to one in which the declining power of labor could be revitalized and reoriented through shared struggle. In a speech before the Illinois AFL-CIO in October 1965, he saw two "deep menaces" facing labor and Black people: automation and the movement of capital to the antilabor South.[61] In this speech, he presented a synoptic view of the intertwined development of automation and the receding power of labor unions:

> The advance of automation is a destructive hurricane whose winds are sweeping away jobs and work standards. The new awareness that America in its glittering prosperity still has nearly forty million poor reveals the dangers facing labor and the unfinished tasks it faces. . . . Where there are millions of poor, organized labor cannot really be secure.
>
> One of the most publicized areas of the poor, Appalachia, is the huge ghost town of the mining industry overcome by automation and new products. In a few years, steel will have lost one-third of the jobs it had in 1950 as new methods and equipment blot out employment. Food packing, auto and electrical assembly, all industries of this state, are visibly scarred by the consuming flames of automation. The process does not abate because it has socially undesirable consequences, but accelerates because it is invariably profitable to industry to shrink jobs.[62]

King then connects this restructuring of capital and labor to the displacement of people, and especially the Black working class, into urban "ghettos" where they either find low-paid, nonunionized service jobs or disappear (at best) into an inadequate, racist, and sexist welfare system. The general economy of the slum was thus not simply an analysis of a geographically delimited Black urban crisis but instead a way of thinking about capitalism in its totality—the relationship between Black and White workers, race and class, and the migration of both labor and capital from the South to the North, then back again. King in this and other speeches notes that capital's emergent return to the South was because the region was anti-union, there were few legal protections for workers, and it was a place where the wage was structured by a "deeply rooted Negro poverty" that "weakens the wage scale there for the white as well as the Negro."[63] King argues here that the cybercultural era is one in which a new iteration of the historic production of racial difference, which capital uses to extract surplus value through low wages and racialized conflict, is met by a new threat of mass unemployment, mass impoverishment, and geographic reorganization, which will affect Black Americans first but which will ultimately wipe out the thirty years' worth of wage increases that primarily Northern unionization struggles had produced. In response, he urged a renewed antipoverty movement at a "time when labor is still strong, and the civil rights movement is dynamic and expanding."[64]

The reorganization of capitalism through automated and geographic value extraction and dispossession returns us to the domestic colony, and through it to Black Power. In King's formulation, the domestic colony is a political value system in which Black people are racialized and differentiated through extractive methods—including so-called race taxes, such as rent increases, high food prices, and discriminatory hiring practices that produce unemployment and drive down all wages—as well as through a policing apparatus that maintains segregation and prevents political action. In *Where Do We Go from Here?*, King productively struggles with the rebellion in the domestic colony of Watts, Los Angeles, which he believed was the result of "white backlash" to the Civil Rights Movement and the rise of Black Power.[65] The "Black Power" chapter of *Where Do We Go from Here?* is especially interesting because he argues with and against the varieties of Black Power as they existed in the brief period between Stokely Carmichael's first call for Black Power in June 1966 and the middle of 1967, when King authored the book. Although King ultimately disidentifies with the slogan (out of fidelity to nonviolent civil disobedience), the chapter is, more than anything, an agreement with Black Power in principle because Black Power, "in its broad and positive meaning, is a call to black people to amass the political and economic strength to achieve their legitimate goals" in the face of White Power.[66]

Although it does not feature explicitly in the "Black Power" chapter of *Where Do We Go from Here?,* the theory of the domestic or internal colony was central to both King's thought and to many from the Black Power movement. The theory of the domestic colony has a long history dating at least to Martin Delany's description in 1852 of Black America as "a nation within a nation," through the Communist International's Black Belt thesis of the 1920s, and to its prominent theorizations by Harold Cruse, Malcolm X, the Revolutionary Action Movement, and the Black Panthers, among others, in the 1960s.[67] The concept was popular enough that Richard Nixon, in a 1968 speech promoting Black capitalism, critiqued the welfare system because he claimed it created a Black "colony within a nation."[68] In Robert Allen's in-depth theorization of domestic colony theory in *Black Awakening in Capitalist America,* he argued that while the theory broadly applied to the history of Black peoples in the United States, by the late 1960s the United States was best characterized by a form of internal neocolonialism. Historic forms of racialized surplus-value extraction, segregation, and political domination still existed. However, he argued that White capitalism was adapting to the Black freedom struggle; through groups like the Ford Foundation, a Black bourgeoisie was being created that would have limited access to political and corporate power and would provide a form of indirect rule to manage the exploitation and domination of the Black masses.

King and Allen's approaches to domestic colonialism differed somewhat.[69] Yet they share a significant concern with the role of automation in the continued (re)production of the domestic colony. Technology in a positive sense was central to what King called the "World House" and "networks of mutuality": communication and travel helped connect the Black struggle in the United States to decolonizing movements around the world.[70] But automation specifically was transforming a domestic colonial underclass into a potentially permanent group of the unemployed. If before automation one of the forms that domestic colonization took was supplying labor power for a variety of “unskilled” and poorly paid jobs, these were increasingly at risk of either disappearing permanently or becoming subject to greater competition from people of all races as more and more jobs disappeared.[71] For this reason, King argued that Black Power was perhaps most “unfortunate” as a political concept because it “gives priority to race precisely at a time when the impact of automation and other forces have made the economic question fundamental for blacks and whites alike. In this context a slogan ‘Power for Poor People’ would be much more appropriate than the slogan ‘Black Power.’”[72] Allen himself believed that Black cultural nationalists and Black capitalists had such limited power that they could never produce equality, or Black “power,” within the capitalist system. Instead, in *Black Awakening,* he advocated for an interracial anticapitalist movement because of the constitutive role of the production of racial difference to the extraction of value; capitalism thus had to be destroyed in order for the Black freedom struggle to be realized. As this suggests, for King, as well as for Allen, automation, Black Power, and the domestic colony were interlinked concepts that they grappled with in order to understand the politics of the emergent cybercultural revolution. As in *The Triple Revolution,* the automation and cybernation revolution was of absolute importance; however, the revolution for King and Allen would best be described as not new but instead continual—continual in the sense that the inequalities of automation and other technologies like mechanization are constitutive, evolving features of the history of capitalism dating to the plantation machine.

Yet whether he discussed automation in terms of crises in Keynesian capitalism, the domestic colony, Black Power, or unionization and wage labor, like the AHC, King focused primarily on automation’s impacts on Black, and White, men. For example, he praised Black Power because of its “psychological call to manhood.”[73] In *Why We Can’t Wait,* his book on the “revolutionary” year of 1963, he wrote that automation

> may be blessings to our economy, but for the Negro they are a curse. Years back, the Negro could boast that 350,000 of his race were employed by the railroads. Today, less than 50,000 work in this area of transportation. This is but a symbol of what has happened in the coal mines, the steel mills, the packing houses,

> in all industries that once employed large numbers of Negroes. The livelihood of millions has dwindled down to a frightening fraction because the unskilled and semiskilled jobs they filled have disappeared under the magic of automation.[74]

According to King, the potential inclusion of Black men into the Fordist–Keynesian family wage and the welfare state was being undermined by automation. As Melinda Cooper argues in *Family Values,* the belief that the welfare state should be expanded to include households headed by Black men was broadly shared by liberals, many on the left, and neoconservatives in the mid-1960s. At the same time, King, like many during this period, argued that welfare provisions shouldn't just be extended but in fact through a guaranteed income be made more expansive, in part to address what he saw as the problem of overaccumulation produced by the success of Fordism's rationalization of production. He wrote in *Where Do We Go from Here?* that a guaranteed income was necessary because "we have come to the point where we must make the nonproducer a consumer or we will find ourselves drowning in a sea of consumer goods. We have so energetically mastered production that we now must give attention to distribution."[75]

In "Gender Trouble: Manhood, Inclusion, and Justice," Shatema Threadcraft and Brandon Terry argue that in order to address the often patriarchal nature of King's politics, it is necessary to "think with King against King."[76] To conclude this chapter, I turn to King's call for a robust UBI and for an interracial Poor People's Campaign to secure equality. These sought to push beyond a Keynesian political economy that attempted to secure stability and effective demand through capital–labor compromise and the family wage, but in its institutionalization of racialized inequality, it also secured domestic colonization and the afterlives of the plantation machine. At a moment when automation was realigning the racialized distribution of inequality, through the Poor People's Campaign and a guaranteed income, King took the first steps in rethinking the ways that racialized social difference was central to the organization of inequality in waged labor. However, he also began to rethink the normative role that family structures and the gendered distribution of waged and unwaged work played in Keynesian state capitalist inequality. In calling for both a guaranteed income disarticulated from work and for "new forms of work that enhance the social good," he sought alternatives to the thingification of neocolonial racial capitalism that is constitutive not just of the exploitation and production of racial difference but also the gendered devaluation of reproductive labor, broadly understood.[77] It is also important to note that despite his evident concern with the impacts of automation, these programs did not revolve around a technophilic fetishization of automated machines liberating humanity. Instead, in a moment of actual and potential automated unemployment, rooted in a history of racialized and gendered (de)valuation of forms of activity and human life, there was an opportunity—and necessity—to rethink and practice alternatives to those very categories.

Welfare, Guaranteed Income, and the Poor People's Campaign

Martin Luther King Jr. and all of the activists, artists, and writers that I discuss in this volume faced (at least) two significant problems in confronting the capitalist (de)valuation of Black life, and life more generally, in the cybercultural era. First, how was capitalist valuation changing? How was it connected to the longer history of racial capitalist valuation of life dating to plantation slavery? Understanding this was not a simple task because it required thinking against the logics of bourgeois and Keynesian economics, as well as with and against the logics of Marxist thought. Both, if less so Marxism, disregarded the contingent relationship between technology and race; they also disregarded the role that the production of difference played in surplus-value extraction and in the ongoing differentiation between waged and unwaged labor and freedom and unfreedom. Second, they faced the problem of how to imagine and practice a new political economy for a postscarcity, potentially automated society, in which equality and freedom from necessity were possible. Yet because automation primarily exacerbated the capitalist logics of devaluing life, and because of the resurgence of White Power in response to the Black freedom struggle (and other concurrent counterculture movements), it was also a moment in which forms of unfreedom and wagelessness seemed more dire than ever before.

On March 31, 1968, the date of "Remaining Awake Through a

Great Revolution," his last Sunday sermon at the National Cathedral in Washington, D.C., the key political struggles through which King sought a new valuation of the human in the face of automation were the guaranteed annual wage and the Poor People's Campaign. King's "pilgrimage" to a "democratic socialism," as Thomas F. Jackson has called it, equal to the cybercultural revolution was conditioned especially by his encounter with Black women activists in the National Welfare Rights Organization (NWRO) as well as his attempt to build an interracial Poor People's Campaign. These transformed his sense of, especially, the gendered nature of capitalist valuations of life and how alternative futures must be in part about creating what Ruth Wilson Gilmore has called abolition geographies.[78] Through these two, we see the way he engaged with what was then relatively mainstream Keynesian support for a welfare system and a guaranteed income while simultaneously pushing into new territory that sought, by engaging with the ways that welfare, the wage, and social relations were mediated by the production of racialized and gendered inequality, to arrive elsewhere.

As I've already noted, *The Triple Revolution* and the NCTAEP, as well as other documents from the time, like the 1967 Kerner Commission report (officially the Report of the National Advisory Commission on Civil Disorders), helped mainstream the idea of a guaranteed income in the United States. It was in fact mainstream enough that on his election to the presidency, Richard Nixon introduced a guaranteed income bill that was debated, and nearly passed, in Congress. Called the Family Assistance Plan (FAP), it was designed to replace the Aid to Families with Dependent Children (or AFDC) program and provide what he called a federal "foundation under the income of every American family with dependent children that cannot care for itself"—alongside, he hoped, gaining political support from Black and White working-class men who wanted access to welfare-state provisions. As Premilla Nadasen argues in *Welfare Warriors: The Welfare Rights Movement in the United States*, FAP, which ultimately did not become law, demonstrates the "consensus" during the period regarding a guaranteed income and consumerism, as well as normative racial and gendered ideas about work and the wage and the central role that the family should play in wealth accumulation and distribution.[79]

Today, whether a guaranteed or universal income is proposed by left-wing proponents of a postlabor automated economy or by Silicon Valley executives who wish to further enmesh people in commodified social relations, UBI is usually presented as a new idea for the twenty-first century. In fact, during the 1960s, UBI, especially as a negative income tax, was supported in one form or another across a wide political spectrum, including Richard Nixon, Milton Friedman, and Democratic Party politicians. What they largely shared, along with a desire to "simplify" the welfare system, was the intention of stimulating effective demand and consumer spending, and the belief that the government needed to intervene in the economy to reduce poverty. (Friedman and Reagan would, of course, disagree with the idea of government intervention in principle, if not necessarily in practice.) What they also largely shared was a belief in the principle of the work ethic and that a guaranteed annual wage should be accompanied by work incentives and sometimes even requirements, and that the normative nuclear family should be at the center of social and economic life.[80]

As the quotations I've included above demonstrate, King himself tended to frame automated unemployment, and unemployment and poverty more generally, in terms of its impacts on men and the family wage, as these were mediated by historic forms of racialized inequality. King shared this gendered normativity and the belief that Black men should have access to welfare provisions not only with Nixon but also with Black capitalists, as well as automation theorists like the LRBW. This was challenged explicitly, and most powerfully, by Black women in the NWRO, who were not automation theorists but whose concerns with racialized poverty would affect King's own thinking on the subject. As Nadasen argues, the NWRO "articulated a version of black empowerment that differed from the widespread patriarchal discourse at the time by rejecting the male-centered solutions to poverty premised on a two-parent, male headed family model" and forced often patriarchal civil rights leaders to "confront problems of poverty more directly" through their demands for a minimal standard of living and dignity in the face of a racist, sexist, and dehumanizing welfare system.[81] This activism took a number of different forms, depending on whether campaigns focused on abuses of individual social workers or on mass demonstrations advocating for not just a more generous and equitable distribution of welfare but also an antiracist and antisexist one. By the time of FAP, a guaranteed income was at the center of their activism and they were influential enough to participate in the congressional debates about the form it should take, and its ultimate demise.

For Black women in the NWRO, like Johnnie Tillmon, a guaranteed income would not only help to solve the "welfare crisis" but also "go a long way toward liberating every woman"; it would acknowledge that "women's work is real work" and that they should be paid "a living wage for doing the work we are already doing—child raising and housekeeping."[82] The Philadelphia Welfare Rights Organization argued that the welfare system of the 1960s was designed to keep "a cheap slave work force on hand . . . to collect its garbage, sweep its streets, harvest its crops, and fight its wars." Alternatively, a guaranteed income "would eliminate the stigma of being a special class or more accurately caste within our society since it is for all families."[83] Thus, as Nadasen argues, for the NWRO and its associated chapters, a guaranteed income would do multiple forms of work: it would recognize care work as real work; it would free women from dependence on men and degrading labor outside of the home; and by acknowledging multiple types of work as work, it would debunk racist characterizations of Black women in particular as lazy and only in search of welfare.

King's political activism around automated unemployment and the NWRO's struggle for the recognition of domestic and care work as work might appear at first to be unrelated. However, Black Power cyberculture asks us to think beyond automation as only a production technology that affects employment and toward automation as a political problem related to the ways capitalism values life and thus to contestations over social reproduction and more-than-capitalist forms of value. For Martin Luther King Jr. and Black Power cyberculture, automation marked a specific crisis in Fordist–Keynesian social reproduction, and the struggle over new forms of value, welfare, and activity thus became political problems. Although far from causal, questions about a racialized history of waged labor brought into crisis by automation thus also provoked questions about the racialized and gendered nature of forms of nonwaged labor, as well as how these categories are conceptualized and how part of the political process of justly distributing society's surplus is also about creating new forms of thought.

The division between racialized and gendered waged and unwaged work is a long-standing political problem. As Saidiya Hartman argues, Black women's reproductive and productive labors have not been easily assimilated into masculine and worker-centric narratives of the "revolutionary" overcoming of racial capitalism in large part because "Black women's labors have not been easy to reckon with conceptually."[84] Further, the continued theft and regulation of Black women's productive and reproductive capacities endure as constitutive features of how capitalism extracts value and how forms of waged and unwaged work are differentially gendered and racialized in the afterlife of slavery. King's own characterization of the reification process of the plantation machine and automation as primarily being about men's unemployment in industry points to the long-standing ways that value has been gendered.

King had, dating to 1964, called for an Economic Bill of Rights for the Disadvantaged that would include a work guarantee, alongside a guaranteed income for those who could not work (among other features). However, his income politics were pushed into new territory by the NWRO. Nadasen recounts a key moment in the encounter between King and the NWRO leading up to the Poor People's Campaign. In Chicago in February 1967, Tillmon, Etta Horn, and other NWRO activists directly confronted King over his lack of knowledge about the welfare movement, as well as the often patriarchal positions he took in relation to it. At the meeting, Horn asked him, "How do you stand on P.L. 90-248?" When King demonstrated that he was unfamiliar with the antiwelfare law, he was further questioned about whether he had been at recent Washington rallies in support of welfare and whether he knew about their own pamphlets on the bill. When he gave unconvincing answers, Tillmon directly critiqued him: "You know Dr. King, if you do not know about these questions, you should say you don't know and then we could go on with the meeting." "Shaken," as Nadasen writes, King conceded: "You are right Miss Tillmon, we do not know."[85]

Like the NWRO, a UBI was central to King's sense of a just social democratic welfare state. Robert Allen, like others in the 1960s, was skeptical that a guaranteed income could be a solution to the automated unemployment of Black Americans because the mainstream proposals that were most likely to be put into law were closest to Milton Friedman's negative income tax. These would provide poverty-level funds while eliminating all other forms of welfare, thus requiring people to further incorporate themselves into a structurally unequal labor market in order to reproduce their existence.[86] King's own UBI proposal at the time of his death, however, demonstrated the extent to which he had been influenced by encounters with the NWRO. For example, a UBI was not just crucial to meeting the automation and effective demand crises but was also part of a reparative process to address the constitutive nature of racial hierarchy to capitalist inequality in the United States. In the buildup to the Poor People's Campaign, he also began to downplay the importance of the family wage in his discussion of automated unemployment, and poverty more generally, and he spoke about the "economic paternalism" of capitalism and the welfare system.[87] As he wrote in *Where Do We Go from Here?*: "If the society changes its concepts by placing the responsibility on its system, not on the individual and guarantees secure employment or a minimum income, dignity will come within reach of all."[88] As Threadcraft and Terry argue, this sense of dignity against systematic inequality was a move to address the constitutive ways that value is defined and produced through racialized and gendered formations in the afterlife of slavery. To this end, given both the undervalued labors of Black peoples in the history of U.S. capitalism and the general affluence of U.S. society that these labors helped to produce, in *Where Do We Go from Here?* King argued that a guaranteed income must do two things for it to ensure both dignity and an end to poverty. It must be "pegged to the median income of society, not at the lowest levels of income," and it "must be dynamic; it must automatically increase as the total social income grows."[89]

Alongside new, dignified, valuations of life, contesting the general economy of the slum also required new social spaces. Central to the Poor People's Campaign was who the subjects of a political movement for new valuations of life and just distribution of society's surplus should be—which is to say, where should power come from? This was a question King grappled with in relation to Black Power. In a period of White reaction to civil and human rights, King largely shared the political and economic critiques of many Black Power activists (if not necessarily the culturalist or capitalist tendencies in the movement).[90] Much like James and Grace Lee Boggs at the same time, King argued that the Black freedom movement was increasingly out of sync with the trajectories of capitalist economic development. It instead needed to focus on antipoverty projects and on creating economies of equality. Perhaps most strikingly, as I wrote above, he called for power to poor people instead of Black Power because automation was radically reorganizing the economy for Black and White peoples alike. The workplace—which, as Frederick Douglass, W. E. B. Du Bois, Lisa Lowe, David Roediger, and others have shown—relies on the production of difference. As King suggests here, automation helps maintain these social divisions by producing racialized forms of struggle over limited jobs, which in turn necessitates new class-based alliances in the struggle over the affluence of an automated society.

In the months after finishing *Where Do We Go from Here: Chaos or Community?*, he would begin preparation for the Poor People's Campaign, which ultimately took place in Washington, D.C., in the summer of 1968. When he introduced the campaign to the Southern Christian Leadership Conference (SCLC) in December 1967, he did so in ways that echoed his geographic and technological critiques of U.S. racial capitalism that I discussed above. He said that the United States was subject to "a kind of social insanity" that was in part the product of geographies of segregation where "affluent Americans are locked into suburbs of physical comfort and mental insecurity; poor Americans are locked inside ghettoes of material privation and spiritual debilitation," and Black and poor peoples could no longer be "placated by the glamour of multi-billion-dollar exploits in space" and "technological wizardry" but "political immorality."[91] This distribution of scarcity in a space age of affluence was not a "value system," he told ministers in the SCLC, where "integration" was possible but rather was one in which "a radical redistribution of power must take place."[92]

Many in the SCLC doubted the idea of a Poor People's Campaign, but for King, a "massive," "interracial," and nonviolent movement of civil disobedience to end poverty was necessary because the Civil Rights Movement had reached its limits, and human rights defined as economic equality required new forms of solidarity.[93] In closing, I want to highlight two features of the movement. First, consistent with the influence of the NWRO and his understanding of automated unemployment, the movement would demand a guaranteed income and a job guarantee. However, as he elaborated in *Where Do We Go from Here?,* jobs should not be defined in terms of increasingly outdated productive or industrial labor but instead of human services. In this alternative to the service industry as it would rise to prominence in the years after King's assassination (and which would absorb many of those displaced from factory work by automation and outsourcing), he defined "human services" as those that were not "capital intensive" and not based on "private enterprise" but instead were jobs "for people from the neighborhoods who can perform important functions for their neighbors."[94] King is developing here the ways that just political economies must be attendant to how often racialized and gendered forms of reproduction, like care work, child-rearing, and education are unpaid and denied forms of work that reproduce labor power for "productive" labor and thus reproduce capitalist inequality. Alternatively, human services must be put at the center of the social reproduction of a human rights–, person-oriented economy and new forms of value, whether this be through a guaranteed income, paid work, or new forms of noncommodified socially necessary activity.

Second, new spatial relations must be created alternative to those of the general economy of the slum and the dialectic between White suburbs where capital was accumulated and Black urban neighborhoods where value was extracted. Racial capitalism relies on segregated spaces, or, as Ruth Wilson Gilmore puts it, forms of "partition and repartition through which racial capitalism perpetuates the means of its own valorization."[95] These include, but are not limited to, racially segregated neighborhoods, the division between spaces of paid labor and unpaid labor, racially and gendered segregated jobs, global production based on logistics technologies, and racialized carceral geographies. Both as King imagined it and as happened in mid-May 1968, caravans began arriving in Washington from Mississippi, San Francisco, Appalachia, and other places around the country. These caravans included a coalition of poor, Black, Chicano, White, and Native American activists. After arrival, they primarily lived in Resurrection City, which was built on the Washington Mall. While it was not supposed to be the point of the Poor People's Campaign but instead the starting point for mass civil disobedience, the founding of Resurrection City paralleled King's description of the relationship between the sanitation workers on strike in Memphis and the politics of reimagining the general economy of the slum: "you are demanding that this city will respect the dignity of labor," whether that be the labor of doctors or of sanitation workers.[96]

While Resurrection City, after an initial period of experimentation in conception, construction, and participatory living, was seen by one of its designers, John Wiebenson, as reproducing many of the limitations of U.S. cities at the time (including forms of segregation),[97] the goal was to create what Ralph Abernathy called the "American Commune" and a "Koinonia, a community of love and brotherhood."[98] This echoes not just King's anticapitalism but the Black social gospel that informed his political project and belief in the need for agape. "Agape," he said at Morehouse on January 1, 1957, "is creative, understanding goodwill for all men. (Yeah) It's a love that seeks nothing in return."[99] For King, such creative love was a necessity for creating new political economies for a post-scarcity age in which automation technologies offered, but were not required for, new social relations beyond "productive labor." But as the general economy of the slum and the Poor People's Campaign show, this does not take place in some imaginary location but instead in spaces that confront the production of racialized and gendered scarcity in a world of affluence. This parallel's Gilmore's conception of abolition geography that "starts from the premise that freedom is a place" that is made through experimentation and people's "social capacities" to organize themselves.[100] It is a practice that is evidenced in postabolition Black communities that sought to destroy the geography of slavery "by mixing their labor with the external world to change the world and thereby themselves—as it were, habitation as nature—even if geometrically speaking they hadn't moved far at all."[101] Especially in the context of the carceral United States that emerged in the wake of the Black freedom movement, Gilmore argues that "abolition geography [must be] capacious (it isn't only by, for, or about Black people) and specific (it's a guide to action for both understanding and rethinking how we combine our labor with each other and the earth). Abolition geography takes feeling and agency to be constitutive of, no less than constrained by, structure."[102]

King's cybercultural praxis was a call to produce new lived experiences of abolition at a moment when Keynesian political economy's ability to reproduce itself was coming undone under the strain of automation and cybernation, White Power resurgence, and anti–social welfare politics.

Examining Martin Luther King Jr.'s automation thought as it necessarily intersected with especially the welfare and antipoverty movements that overlapped with the civil rights and Black Power movements helps open up some of the limits and possibilities of the post–World War II welfare state in the United States. In many respects, King struggled for Black incorporation into the U.S. society of affluence. For example, he sought to expand the power of Black people as consumers, whether through work or a guaranteed income, at a time when the consumer was becoming the primary social subject in the United States. Yet as the Poor People's Campaign, his increasing engagement with the NWRO, and the problems of automated unemployment demonstrate, King was also pushing beyond these limits. The history of racialized and gendered inequality that organized differentiated forms of waged and unwaged labor, freedom and unfreedom, was also the foundation of post–World War II United States affluence. Because of this, King's politics had to push toward economies that did not reproduce these constitutive inequalities. Finally, King's automation politics were not technodeterminist. Instead, the specter of automated unemployment occasioned not only a reconsideration of the issues above but also ethical-political questions about new forms of solidarity, equality, and life that need not be, perhaps paradoxically, premised on robotic, technophilic futures.

2

Initiating a New Plateau

AUTOMATION, POLITICAL ECOLOGY, AND ACTIVITY IN THE WORK OF JAMES AND GRACE LEE BOGGS

In *The Next American Revolution: Sustainable Activism for the Twenty-First Century,* Grace Lee Boggs wrote that it "has also been my good fortune to have lived long enough to witness the death blow dealt to the illusion that unceasing technological innovations and economic growth can guarantee happiness and security to the citizens of our planet's only superpower."[1] It is certainly debatable whether in the early twenty-first century happiness has been delinked from technological innovation and economic growth, at least in the eyes of most Americans. However, Grace Lee and James Boggs have been central to the fact that these have been questioned in activist movements since the 1960s.[2] In the early 1970s, they characterized capitalism in the United States as a system based on the "cancerous growth of production for its own sake" that required racialized and gendered exploitation and for nature and people around the world to be put to work so that (primarily wealthy) Americans could live lives in which value was defined in consumerist terms.[3] In response to this unprecedented, technologically advanced, and postscarcity world of organized inequality, they tried to create a Black Power vanguard movement that would not only replace capitalism with a classless society that more justly distributed social surplus and addressed historic inequalities and environmental destruction but that would also produce a more active, participatory society in which people were conscious of the ways that they were dialectically related to not just other people but to the world in general. One of my primary goals in this chapter is to demonstrate the extent to which the Boggses' ecological politics dates to the cybercultural era. Their work from that period offers a necessary approach to thinking about the mutual constitution of ecological problems and automation, which today, just as during the cybercultural era, are rarely thought together, as well as what political strategies can best organize a postscarcity, classless society that is based on new forms of value.

James Boggs was one of the many authors of *The Triple Revolution* that I discussed in chapter 1. However, alongside economist Robert Theobald, Boggs dissented from the AHC's primarily New Deal–oriented recommendations. He believed *The Triple Revolution*'s recommendations accounted for neither the extent of the technological changes to labor nor the fact that the United States in the 1960s was a postscarcity society in which new ways of valuing "work" and "leisure" were necessary.[4] This chapter will examine in greater depth what I began to do in the introduction: to analyze the Boggses' sense of "scientific" Black Power. They used this concept to argue that the United States was undergoing a moment of racial and class recomposition in relation to automation technologies—a moment also due to the Black freedom and anticolonial struggles that were emerging as new political forces in the wake of the failure of the post–World War II union movement to take real power. At the same time, they argued that Black Power, as an avant-garde, must take on the responsibility of creating new human values and forms of activity, a project they called dialectical humanism.

Here I also examine another aspect of their thought, which I will further develop in relation to Noah Purifoy in chapter 3. For the Boggses, automation was not just a problem of unemployment and work; it was also a political-ecological problem that necessitated new forms of value and social reproduction. Capitalism, as they argued, was necessarily a growth-oriented system that required not only racialized and gendered inequality in the labor process but also required that nature be put to work in destructive ways and for the waste produced by capitalist production and consumption to be unequally distributed. The ecological aspects of automated overproduction were not part of James Boggs's dissent from *The Triple Revolution,* but there is nothing to suggest that the ecological problems of production and consumption would be altered by a new science of political economy as that document conceived it. In fact, cybernated production can be read as potentially increasing these problems because they would result, according to *The Triple Revolu-*

tion, in "a basic reordering of man's relationship to his environment; and a dramatic increase in total available and potential energy"—which is to say that the automated rationalization of production offered the possibility of producing even more commodities and thus, as we know even better today, further planetary damage and unequal impacts on poor and racialized communities.[5]

In "Political Ecology, Distributional Conflicts, and Economic Incommensurability," Joan Martinez-Alier argues that during the 1960s and 1970s, Marxists were skeptical of the Malthusian underpinnings of the period's narratives of collapse, like the Club of Rome's 1972 study *The Limits to Growth,* and thus "chose to interpret environmentalism as either a dangerous anti-industrial romanticism or an upper-class frivolous fad."[6] In doing so, they forgot "that the growth of the economy implies the consumption of more resources and the production of more waste, and so failed to understand either the class content of [environmental] struggles against the effluence of affluence . . . or the environmentalism of the poor."[7] Today this has begun to change. However, too often discussions about ecology, racialization, automation, and capital accumulation remain disconnected, and thus so too do political responses to them. During the cybercultural era, the Boggses argued that the major challenges facing Black—and indeed all—revolutionaries was to help people to see these connections, and by extension challenge them to think deeply about what it means to be human and in what new ethical and creative ways humans might live.

These challenges remain with us today, and returning to James and Grace Lee Boggs's scientific Black Power can help us to meet them. This chapter will primarily examine their work from the early 1960s to the mid-1970s and their historically grounded theorization of the relationship between automation, racialized value extraction, urban land, and hierarchical distributions of the human. Most significant, what I am calling their political ecology was about distribution conflicts and how to create a classless postscarcity economy that equitably (re)distributed society's surplus while contending with a consumer capitalist growth economy that was, and continues to be, based on racialized and gendered forms of inequality and the production of waste and toxicity. To do this, I analyze major books like 1963's *The American Revolution: Pages from a Negro Worker's Notebook* and 1974's *Revolution and Evolution in the Twentieth*

Century, as well as numerous essays, pamphlets, and speeches from the 1960s.[8] The Boggses had a systematic knowledge of Marx, philosophy, and political economy; however these are primarily "movement" texts written as part of continually developing political struggles, and thus their own thought was always adapting in dialectical relationship to those struggles.[9] In this respect, they are works of "low theory," a concept coined by Jack Halberstam that draws on Stuart Hall and Antonio Gramsci.[10] Low theory is counter-hegemonic thought. As McKenzie Wark puts it, "You could think of low theory as what organic intellectuals do. It's defined by who does it and why, rather than by any particular cognitive style. . . . Concepts get formed differently and are meant to do different things when you are trying to think through your own action in the world rather than when you are a scholar of action in the world."[11] To draw out the Boggses' political ecology is to read across their body of work, both for their general conceptual and ethical positions and also for how they conceived of and responded politically to the (continually developing) cybercultural conjuncture.

The question of value was at the center of their best-known text, *The American Revolution,* as it is in all of their lifework, which continued until James's death in 1993 and Grace Lee's in 2015. By raising the question of value, James Boggs was in part concerned with how capitalist technologies of (especially) the factory were organized to extract surplus value. But they were also concerned with value in a much broader sense. In industrial production, automation involved a reorganization of the labor process. As Harry Braverman argues, in a capitalist system where management controls the relations of production, the goal of automation is to de-skill workers and increase management's own power, which is to say that automation is a political intervention to limit worker's agency. James Boggs was an autoworker for most of his adult life, and he was politically formed, in part, through shop floor organizing. This personal history is reflected in early passages of *The American Revolution* where he discusses the "rich" political and economic experiences of workers who were involved in the labor struggles of the 1930s and 1940s. These struggles resulted in an "educat[ion] in the meaning and nature of modern production."[12] By this, he meant that through activism, they had both learned how to organize, and thus potentially reorganize, the relations of production in the factory; but they had also learned a lesson of betrayal when union leaders collaborated with management and the government to reestablish capitalist control after World War II.

This was a loss of values for the labor movement that at the same time spoke to a larger problem of loss and value akin to what Bernard Stiegler has called "proletarianization," or the technologically induced loss of *savoir faire* and *savoir vivre* in everyday life.[13] Along with the reorganization of the labor process, the Boggses believed automation was also leading to under- and unemployment and the creation of a new class of outsiders, largely but certainly not exclusively Black Americans who could no longer find, as I wrote in the introduction, a place inside capitalist production. This meant that not just new forms of racial and class composition needed to be devised; it would also require a new politics focused on creating new postlabor values. As James wrote, "Now that man is being eliminated from the productive process, a new standard of value must be found. This can only be man's value as a human being."[14] The question of value remained at the forefront of James and Grace Lee Boggs's constantly evolving, dialectical thought. As James Boggs asked in the late 1980s, "How do we create a culture that is life affirming rather than life destroying, which is based on caring and compassion rather than on the philosophy of the 'survival of the fittest'?"[15] During the cybercultural era, at the center of this life-affirming politics was an attempt to politically reorient life away from "labor" and toward what they called "socially necessary activity." Through this category, they sought new practices of care and social reproduction that would respect the "labors of past generations" as well as the ways that Black labor in particular had produced a world of abundance that could allow people to work less and move away from a life premised on material overconsumption. At the same time, these practices sought to take account of capitalism as a totality that involved waged and unwaged labor, extraction, production, consumption, and waste, and that necessitated building a classless society out of cities that in the 1960s were both increasingly Black but also increasingly sites of environmental degradation.

This chapter explores three interconnected aspects of their political thought. First, I continue the examination I began in the introduction of their theory of automation and cyberculture. In *The American Revolution,* "The City Is the Black Man's Land," along with

"The Negro and Cybernation" and other essays, they argued that the United States had undergone moments of Indigenous land dispossession, agricultural mechanization, and industrialization, and it was "on the eve of the cybercultural revolution."[16] Each of these was a moment of technological change and new compositions of race and class. However, the 1960s was an especially revolutionary moment, and they called their method of analyzing the history of the present scientific Black Power. At the same time, they used this term to describe their political project. In opposition to cultural nationalist and capitalist forms of Black Power, they argued that as the most exploited members of American and global society, Black people were positioned to lead the struggle for a classless society.

The second half of this chapter turns explicitly to their political ecology. Beginning with "The City Is the Black Man's Land" and continuing into *Revolution and Evolution in the Twentieth Century,* the Boggses argued that automation was not only producing racial and class inequality, but, through increases in production and consumption, was intensifying ecological problems, both on a planetary level and through the distribution of pollution, toxicity, and waste in Black urban communities. The Boggses' analysis of this problem linked planetary damage and racialized inequality; they saw Black urban space as akin to what Anna Lowenhaupt Tsing calls third nature, or that which comes after capitalist destruction of the environment.[17] Because of this, the Boggses argued that the American Revolution must start from "land" in Black cities.

In the third part of the chapter, I look at how a politics that would both remediate environmentally damaged communities and address the dispossession of Black peoples since slavery would also have to be, in today's language, a degrowth project. They argued that the American Revolution would require people to give up consumerism in order to gain a postscarcity, classless society. They called this philosophy of new values dialectical humanism. In this section, I detail the ways shifts in capitalist production also required a shift from a society based on socially necessary labor in production to one based on socially necessary activities and new forms of social reproduction. Analyzing dialectical humanism's features will also lead into chapter 3 and Noah Purifoy's attempt to create new forms of value through community art in what he called the materialistic world.

The Cybercultural Revolution and Scientific Black Power

In *The American Revolution,* James Boggs writes, "If you can once get the average American to stop blaming everything on the communists (or the Negroes, or the Jews, or the Italians)," then they will say that the biggest problem facing the United States in the 1960s is the problem of automation.[18] However, the average American will say this with "a distant look in his eyes," believing that whatever the problem of automation might be, it will soon pass away, having no impact on his everyday life. Writing in 1963, Boggs strongly disagreed with this. However, on the one hand, in some respects, we would be forced to agree with this average American: the future that James imagined did not come to pass, and the outsiders who Boggs argued were being produced by automated unemployment were soon absorbed into the service industries, an expanded carceral system, and new forms of debt. On the other hand, these examples point to the ways that James's analysis of automation in *The American Revolution* was prescient. What he argued most extensively was that automation was producing major changes to the technical composition of labor, both in the factory and more generally, as well as in the composition of race and class, which was undoubtedly true.

The American Revolution is a wide-ranging text, which includes analyses of the Cold War and of the state of U.S. imperialism. At its center are the revolutions in production, civil rights, and the emergence of Black Power, as well as the "fears," "prejudices," and "outmoded ideas" that prevent everyone in American life, and especially Marxists, "from grappling with the new realities of our age."[19] The outmoded idea that James begins with is that of class and its relationship to race. As he writes on the very first page, in "America, more than in any other country, the revolutions in the mode of production have been accompanied by changes in the composition and status of classes."[20] Initially, *The American Revolution* was printed as the sole piece in the July–August 1963 edition of *Monthly Review;* it would eventually be translated into

Italian and printed in *Classe Operaia*. Alongside work by C. L. R. James, Raya Dunayevskaya, and Grace Lee Boggs as the Johnson-Forest Tendency, Martin Glaberman, Charles Denby, and the LRBW, as well as publications like *The American Worker, News and Letters,* and *Correspondence,* it pioneered class composition theory and heavily influenced the Italian *operaismo* and autonomist movements, as well as Herbert Marcuse's *One-Dimensional Man.*[21] In it, the status of class is a political problem in many senses, and in this first moment, James acknowledges the confusion brought about by rapid changes in the mode of production. These included generational shifts in labor, especially from agriculture to industrial, technical, and office work, as well as new forms of education, and the introduction of women and Black people into offices and factory work (especially because of World War II).

The ramifications of these are extensive. First, categories like the "working class" no longer hold, and no class can now be considered a "homogeneous segregated bloc."[22] The lack of homogeneity is reflected inside the factory and in the increasing number of technicians and middle-class workers. In the 1930s, "workers were divided by race and nationality prejudices ('Dagos,' 'Wops,' 'Polacks,' 'Niggers,' 'Buffaloes,' etc.)."[23] This resonates with Cedric Robinson's sense of racial capitalism defined as the ways that racialism, or the exploitation and (re)production of social difference, is a material force that divides workers in order to better exploit them to extract surplus value. The role of racial difference in so-called free labor is a problem that Frederick Douglass encountered on the docks of New Bedford after his exodus from slavery and one that, following David Roediger, has often been called the wages of whiteness.[24] However, for James Boggs, while "the native-born white" is the (no longer relevant) normative subject of both the working class and its revolutionary potential, he argues here that the introduction of a racially and ethnically diverse work force meant that "fights" over the labor process complicated the apparent homogeneity of whiteness. At the same time, he argues that the factory was also a "social melting pot" and that through CIO, worker-led organizing around World War II, "the workers created inside the plants a life and a form of sociability higher than has ever been achieved by man in industrial society."[25] Thus, the material force of social difference could not only be overcome through organizing but could also be the testing ground for a classless society and new forms of social life.

In order to understand the new heterogeneity of class composition in the 1960s, Boggs argues it is important to start from this premise: the highest stages of organizing and social life, whose agents were now mostly a "vanishing herd," despite numerous wildcat strikes in the 1950s.[26] This disappearance was for multiple reasons. First, the CIO failed to take control of the means of production and create new relations inside the factory and in society at large. Instead, union leaders became a bureaucracy and turned power over to management, in collusion with the government, after the war, thus undoing those new forms of sociality. Second, even in the early 1960s, factory work was less and less central to employment, thus undermining old assumptions about where the working class was located. Third, automation was changing the technical composition of labor in the factory itself. Automation was not only increasing unemployment but also creating a new binary division of labor at the point of production. On the one hand, in factories, there were those workers who still primarily worked with, or were subject to, the assembly line, who in the 1960s in Detroit were becoming majority Black. On the other hand, there were what he called "new workers" who were "part and parcel of the new process of production." These were entirely nonpoliticized workers whose "ideas are so crucial to the direction of the work that they are inseparable from management and the organization of the work," and they controlled "the flow of production itself."[27] Boggs described these workers as not like the "old inventor-geniuses"[28] that management would bring in to problem solve on the assembly line, but instead were much more akin to embodiments of what Marx famously called in the *Grundrisse* the "general intellect." In the "Fragment on Machines," Marx argued that even in the mid-nineteenth century, machinery and fixed capital had developed to the point that surplus value wasn't solely extracted from living labor at the point of production but instead from "general social knowledge" becoming a "direct force of production."[29]

This new technical organization of labor wherein the factory was characterized by an antagonism between an increasingly small number of assembly-line workers and a small but increasingly significant number of new knowledge workers marked one facet of the new class composition of the cybercultural era. What he argued was the most significant, but still emergent, form of antagonism was between employed workers in any industry or line of work and the increasing number of unemployed workers or outsiders produced by automation. Thus, in the early pages of *The American Revolution,* Boggs charts how the cybercultural revolution is a dynamic reorganization of labor in which old assumptions about capital and the working class no longer held. The challenge facing revolutionaries is to be attendant to this dynamism and heterogeneity without falling back on old categories of knowledge that are now politically useless.

Alongside class, the second feature of this new heterogeneity was that of race. I have already discussed in the introduction how in "The Negro and Cybernation" James, like King, argued that automation was the most recent moment in a history in which technologies like the cotton gin during plantation slavery and farm mechanization in the early twentieth century were part of changes to the racial composition of labor. However, as political technologies put to work at particular conjunctures, they had different impacts. The cotton gin intensified the need for Black labor; alternatively, farm mechanization put Southern agricultural workers out of work, sending them to Northern cities. There, especially because of New Deal programs and World War II, they had found work in factories. However, automation, while an historical product of these relationships, was unprecedented because it seemed to be ushering in a new form of racial and class composition whereby Black Americans had "nowhere to go" and would become the first wave of a new generalized mass unemployment.[30] Thus, the cybercultural revolution had the potential to either reinforce existing forms of racial division or alternatively to potentially undo them and replace them with new forms of social difference, especially between the employed and unemployed, less tethered to race than ever before. In this respect, "automation is the stage of production that carries the contradictions of capitalism to their furthest extreme," sharpening antagonisms and increasing the importance of political struggles between those trying to advance new forms of freedom and those opposed to them.[31]

This new class antagonism, between those with jobs and those without them, meant that the labor movement could no longer be the unquestioned agent of political development. The (at least initially) highly racialized nature of unemployment meant two divergent things. On the one hand, it meant that the Civil Rights Movement's quest for integration into a capitalism in crisis was misplaced. However, on the other hand, James Boggs argued that introducing race into discussions of automated unemployment and poverty politicized them and removed the veneer of automation as a universal, determinist technology. As he said, race made automation "social because it poses the relations of man to man."[32] Race and the Black freedom movement thus requires that one historicize the development of automation as a political technology even as it simultaneously alters the struggle's terrain, moving it from solely one of jobs and the nature of work inside the factory and the office to the social relations of society at large. For James and Grace Lee Boggs, this meant specifically that the Black freedom struggle must necessarily become an anticapitalist one for a classless society and new social relations.

This dynamic was one that "American Marxists" had entirely missed because they were trapped into thinking of Black Americans strictly as a "race" and not as the "most oppressed and submerged sections of the workers, on whom has fallen most sharply the burden of unemployment due to automation."[33] In *The American Revolution*, James Boggs confronted the failure of the CIO and American Marxists to even attempt to take power and create a classless society. At the same time, it is a text that begins to confront what would soon be called Black Power. In contrast to civil rights, the integrationist aims of which the Boggses believed were out of sync with the cybercultural revolution, they saw the emergent class of Black outsiders as a potentially revolutionary force. They associated this movement with people like Robert Williams, Malcolm X, and the Nation of Islam, all of whom had tremendous influence on the young Black urban masses. Over the rest of the decade they would, however, contrast their own political project to that of the Nation of Islam and other cultural nationalist and capitalist forms of Black Power, as well as against most varieties of White radicalism. They called this political-theoretical movement a "scientific concept" of Black Power.

My discussion of *The American Revolution* should already give a clear sense of their scientific concept. However, they make it explicit in their 1967 essay "Black Power: A Scientific Concept Whose Time Has Come," from which I quote at length to give a full sense of the contradictions between race, technology, economy, and politics posed by the cybercultural conjuncture:

> Those progressives who are honestly confused by the concept of Black Power are in this state of confusion because they have not scientifically evaluated the present stage of historical development in relation to the stage of historical development when Marx projected the concept of workers' power vs. capitalist power. Yesterday the concept of workers' power expressed the revolutionary social force of the working class organized inside the process of capitalist production. Today the concept of Black Power expresses the new revolutionary social force. . . . The uniqueness of Black Power stems from the specific historical development of the United States. It has nothing to do with any special moral virtue in being black, as some black nationalists seem to think. . . . The development of technology in the United States has made it impossible for blacks to achieve economic power by the old means of capitalist development. The ability of capitalists today to produce in abundance not only makes competition with them on an economic capitalist basis absurd but has already brought the United States technologically to the threshold of a society where each can have according to his needs. Thus black political power, coming at this juncture in the economically advanced United States, is the key not only to black liberation but to the introduction of a new society to emancipate economically the masses of the people in general. For black political power will have to decide on the kind of economy and the aims and direction of the economy for the people.[34]

The notion of a "science" of Marxism dates to Friedrich Engels's work in the 1870s and 1880s. In *Socialism: Utopian and Scientific,* Engels contrasted the utopian socialism conceived by the "ingenious brain" of people like Saint-Simon, Fourier, and Robert Owen to a Marxist concept that developed out of the discovery of the theory of surplus value and the historical development of class struggle.[35] For Engels, this political history was then confirmed by nature itself, which, he argued, was a continually evolving dialectical process. After Engels, Marxist science took many different forms over the course of the next century, from that of the Second International and the German SDP to Mikhail Bukharin and the dialectical materialism of the Soviet Union.[36] Because of its association with the Soviet Union, scientific socialism was on the wane by the time of the Boggses' development of scientific Black Power, with a major exception being Louis Althusser's renovation of what he called Marx's "science of the history of social formations."[37]

Given this largely unfashionable history, we might see Grace Lee and James Boggs's thought as outmoded. Instead, the above passage shows their sense of scientific Black Power to be a counterdiscourse that critiques the assumptions of the Marxist scientific-dialectical tradition while opening up new avenues of political thought out of it.[38] Introducing race not only socializes automation; it also socializes both Black Power and the dialectic. In some respects, automation and cybernation coincide with Engels's sense of scientific Marxism: it is the result of the history of capitalist class antagonism, although here class is, of course, dialectically related to race. The cybercultural revolution is the result of these multiple antagonisms; however, so is Black Power and the racial category "black." Far from being a cultural or racially essentialist category, blackness is, in its first moment, the result of what they call elsewhere the technocapitalist "undevelopment" of Black peoples.[39] As they note in "Manifesto for a Black Revolutionary Party," automation and cybernation is a double-edged sword; it has made Black youth in particular "expendable to the economy." However, at the same time, although now rendered "obsolete," automation has "also liberated blacks for the first time in their history on this continent from the necessity to work on behalf of white development."[40] Two possibilities thus emerge out of this obsolescence: deepening unfreedom or new politicized forms of freedom.

In the urban rebellions of the 1960s, they saw a socially aware yet unorganized mass anticapitalist movement of outsiders who knew that White Power was organized to deny them full access to the benefits of space-age affluence. The Boggses argued through their notion of scientific Black Power that the historical development of racial capitalism meant that Black peoples were the avant-garde of the struggle for a classless society because they had been the most undeveloped—the most subject to the "wear and tear of constant use"—by capitalism.[41] However, unlike integrationist politics and

Black capitalism, this meant there was a special responsibility to be attentive to the historical forces that produced racial difference in the cybercultural era. True Black freedom could only be had by "resolv[ing] the problems of society as a whole and not just those of black people. In other words, Black Power cannot evade tackling all the problems of this society because at the root of all the problems of black people is the same structure and the same system that is at the root of all the problems of all the people."[42] The second task of scientific Black Power was thus to form a revolutionary party that would keep the struggle for a classless society at the forefront of the mass struggle of Black outsiders. In "Manifesto for a Black Revolutionary Party," *Revolution and Evolution in the Twentieth Century,* and in other essays, they argued that the development of technocapitalism in the United States meant that the American Revolution would be the first in history to not be about "economics" but instead "politics." It would have to be a revolution that did not simply utilize technology, expropriated from capitalist control, to free all people from unnecessary labor, but one that would more broadly create a society in which all people actively participated in creating a world based on well-being and not exploitation and oppression. Doing this would necessarily require the creation of new values, and they emphasized that "revolutions" were not rebellions of instantaneous change but instead long processes in which the best of historical struggles for freedom would be drawn on to "initiate a new plateau" for humanity.[43] Because capitalist exploitation was an imperial and colonial process, the American Revolution, while specific because of the extent of its overdevelopment, would also necessarily be part of a global struggle. In "The City Is the Black Man's Land," they emphasized that by "black," they meant a "political designation to refer not only to political Afro-Americans but to people of color who are engaged in revolutionary struggle in the United States and all over the world."[44] However, as the title "The City Is the Black Man's Land" suggests, the terrain from which they would begin this political project was that of land in cities that were increasingly Black, but also increasingly reorganized in the aftermath of Fordist automated capitalism where the ecological effects of that mode of capital accumulation were increasingly acute, especially in terms of racialized exposure to waste and toxicity.

"The City Is the Black Man's Land"

The Boggses' focus on the urban "land issue," as they called it in *Revolution and Evolution in the Twentieth Century,* is most often attributed to their post-1960s community work in Detroit.[45] With the waning of the Black Power and other revolutionary movements, the Boggses became particularly invested in community and environmental justice, in Detroit above all.[46] In *The Next American Revolution,* for example, Grace Lee Boggs placed a great deal of emphasis on the relationship between environmental justice and urban agriculture—an emphasis she shared with other Detroit-based groups, such as the Gardening Angels.[47] There, she praised Rebecca Solnit's article, "Detroit Arcadia: Exploring the Post-American Landscape," for how it looked beyond the visual phenomena of postindustrial decay to both the histories of that undevelopment and also to potential futures that reimagine urban space. Solnit is one of today's most vital popular voices critiquing neoliberalism's urban and environmental impacts. However, despite Grace Lee's praise, Solnit's claims about urban agriculture in Detroit actually sheds light on how the Boggses' work offers an alternative history of urban ecology. In the essay, Solnit writes that despite its many important features, Detroit's "experiment in utopian post-urbanism . . . appears to be uncomfortably similar to the sharecropping past their parents and grandparents sought to escape."[48] In the 1960s and today, the Boggses did argue that the land issue is connected to the history of sharecropping, especially in light of the ways that today's "urban homesteading" movements often repeat histories of racialized settler colonial dispossession. I argue that the Boggses' political ecology actually dates to the cybercultural era and that their analysis in the period (as well as throughout their lifework) demonstrates that scientific Black Power's sense of urban land did not repeat the history of sharecropping but instead was a political practice that sought alternatives to how capitalism put human and extrahuman nature to work and extended the rationalities of Indigenous dispossession and the plantation. Further, it was an attempt to practice an alternative to the current form of the human that the Boggses understood to be a "dehumanized" figure of material overconsumption and economic growth—a form that conceptualizes rights only as the "right to exploit" nature and other people. As they note in

Revolution and Evolution, "Having dehumanized themselves by exploiting the natural resources of the whole world and destroying whole peoples in the pursuit of their own material comforts, the American people are now confronted with reality"—a new reality of resource pressures, racial hate turned inward and abroad, and increasing evidence that "these problems cannot be solved by technology, any more than the war in Indochina could be won by technology."[49]

As this remark suggests, the Boggses' political ecology as it developed over the course of the 1960s was motivated by understanding three interrelated problems: the war in Vietnam as a technological war, and my focuses in this chapter: the automated overproduction of commodities that produced a new class of outsiders; and their examination of capitalism as a systematic process of extraction, production, consumption, and waste that used Black and poor communities as a surplus population and a waste sink while also producing a condition of metabolic rift in the population at large. This mode of thought reached its culmination in 1974's *Revolution and Evolution,* but it initially begins with their struggle to understand automation as a political technology.

The cybercultural conjuncture was producing new forms of automated precarity that seemed to be reorganizing race and class. In "The American Revolution: Putting Politics in Command," they claimed that the result of "rapid technological development and the concentration of skills and power at one pole has been a systematic and continuous decline in skills, responsibility, and participation at the other pole, particularly at the very bottom of the ladder among the blacks, Mexican Americans, Puerto Ricans, Indians, etc." Indeed, technological development was, and is, always connected to multiple and potentially conflicting forms of racialized proletarianization and (attempts at) dispossession—not only of skills in the workplace but of all life-making capacities, rendering people dependent on the commodity form without full access to it.[50]

As I have begun to argue, for them, as for King, automation was also part of a history of land use and agricultural technologies, like mechanization, in which racialized forms of proletarianization and dispossession were connected to a systematic reorganization of the geographies of capital accumulation. In "Black Power: A Scientific Concept Whose Time Has Come," they provided a condensation of this history: "In the twentieth century the United States has advanced rapidly from a semiurban, semirural society into an overwhelmingly urban society. The farms that at the beginning of the century still employed nearly half the working population have now become so mechanized that the great majority of those who formerly worked on the land have had to move into the cities. . . . The city is now the black man's land, and the city is also the place where the nation's most critical problems are concentrated."[51] Now, with no where else to go, primarily Black Americans were outsiders on urban land, where they comprised a devalued surplus population more than a reserve labor army because as they saw it, automation was reducing employment possibilities. Here the connection between automation and mechanization as political economic machines for devaluing life is made apparent. They also begin to draw out broader ecological implications, comparing Black urban communities to Appalachian communities—not because they had never been developed but instead just the opposite:

> Scientifically speaking, the physical undevelopment of the black community is decay. Black communities are used communities, the end result and the aftermath of rapid economic development. . . . The physical structure of black communities is like that of the abandoned mining communities in Appalachia whose original reason for existence has been destroyed by the discovery of new forms of energy or whose coal veins were exhausted by decades of mining. It would be sheer folly and naivete to propose reopening these mines and starting the process of getting energy from their coal all over again. When one form of production has been rendered obsolete and a community devastated by an earlier form of capitalist exploitation it would be supporting a superstition to propose its rehabilitation by a repetition of the past. You don't hear any proposals for white capitalism in Appalachia, do you?[52]

Looking at this passage from the vantage point of the early twenty-first century is jarring because this is in fact what has happened in the (at least rhetorical) attempts to rejuvenate the white capitalist coal economy as part of a new racialized antienvironmental politics. What the Boggses called this "naivete" is in fact a recurring feature of capitalism, which, as Jason Moore argues in *Capitalism in the Web of Life,* is both crisis driven and "a way of organizing nature" where accumulation is based on producing human and extrahuman nature as "cheap."[53] Even though they did not anticipate today's specific "rehabilitation" of "earlier forms" of fossil capitalism, the Boggses were prescient to see these two different forms of undevelopment—Black urban and White Appalachian—as part of capitalism's politics of continually traversing new commodity frontiers, and in doing so relying on "cheap," racialized human and extrahuman nature to do so both in production and disposal. As a moment in the cybercultural revolution, automation's specific effects might vary: unemployment, intensified labor, predatory inclusion, incarceration. However, the logic of automation and mechanization as a political project is one of displacement, precarity, and modes of accumulating capital that continually seek to devalue work and life.[54]

In the United States, the Boggses argued that this racialized devaluation of human and more-than-human nature, as well as work, culminating in cybercultural dispossession, Black urban political crises, and environmental damage, was founded on the Puritan dispossession of Native peoples. In their writing during the period, the Boggses did not make overt political connections between Black Power and "Red Power," despite the extent to which the American Indian Movement was an urban one that developed partly in response to the United States government's post–World War II termination policies after the passage of House Resolution 108 in 1953.[55] The Boggses primarily saw Native peoples as those dispossessed and displaced onto reservations instead of as an increasingly large part of cities like Los Angeles and San Francisco.[56] More significant, their theorization of urban land, like their domestic colony theory more generally, largely left unexamined how then-contemporary struggles, exemplified by the occupations of Alcatraz, the Bureau of Indian Affairs headquarters in Washington, D.C., and of Wounded Knee, were part of long-term movements by Native peoples for forms of sovereignty, self-determination, and land that, because of the specificities of settler colonialism in the United States (land dispossession and African slavery), both conflicted with and complemented their own. However, despite their specific focus on a Black politics of urban land, their thought resonates with contemporary work on settler colonialism and capitalism in that they argued that settler dispossession of land was not simply a form of capital accumulation but also a radical reordering of social relationships and conceptions of life that were the precondition for technocapitalist devaluation of life and work in the late twentieth century. For them, Puritan dispossession of land involved a conception of "European men/women" based on the enclosure and privatization of land—a conception too limited to include other conceptions of "humankind" and relationships to land.[57]

This conflation of land use and what constitutes a proper human was then extended in the development of industrial modernity that they, like C. L. R. James, saw as modeled on the slave plantation's organization of labor, in which diasporic African peoples in the Americas were dehumanized as commodities and labor power, in the process becoming the "first working class on the land."[58] The intertwined history of racialized labor and the domination of the nature had broad implications on the then-present. Industrial capitalism was, and is, geared entirely to overproduction. With post–World War II innovations in forced obsolescence, Americans, the Boggses argued, were increasingly becoming subjects of a "throwaway mentality" that bound them more strongly to forms of consumption in an (impossible) attempt to satisfy what they falsely believed were their needs while condemning a large percentage of the planet to labor in extraction and production industries in order to produce those commodities. This overproduction, increasingly rationalized by automation, is then transformed and redistributed, becoming forms of pollution and waste destroying the planet as a whole, but especially, for the Boggses, the urban spaces that Black Americans had been recently displaced into. In *Revolution and Evolution in the Twentieth Century,* they called this, as well as the potential for political alternatives, "A Unique Stage in Human Development."[59]

As I noted at the beginning of this chapter, Marxists during this period did not always take seriously the ecological impacts of capitalism. However, within counterculture movements, whole-system ecology based on theories of feedback, which sought to connect extraction and consumption to environmental degradation, was on the rise.[60] What separates the Boggses from this version of ecological theory—in addition to their commitment to dialectics and not feedback (although they were certainly conversant in cybernetic thought)—was that their investigation of automation as a political technology that devalued human life led them to see how race, and social difference more generally, was constitutive of how techno-capitalism exploited, appropriated, and produced nature as a cheap commodity (or potential commodity), and how the reproduction of racial and class inequality was then linked to the ways capitalism externalized, both physically and economically, waste disposal. In this respect, they anticipated what Françoise Vergès calls the "racial Capitalocene."[61]

Vergès uses this concept to examine how forms of racialization are central to the distributions of inequality in our recently named geologic epoch, the Anthropocene. In doing so, she argues that there is no longer a "natural" world but instead a highly technological "Promethean" one in which a transcendent "Man" is believed to be capable of inventing a "technical solution to any problem." However, it is also a world where abundance, planetary transformation, and technological possibilities are premised not only on racialized forms of capital accumulation but also on racialized distributions of waste and pollution.[62] For the Boggses, "the black man's land" is a thoroughly technological space where the formerly enslaved and colonized people who helped to build capitalist modernity, and who have been most intensely subject to its project of commodifying life (whether through exploitation or appropriation), are then distributed in relation to industrial waste and toxicity. Racialization and waste production are the coconstitutive central dynamic of the racial Capitalocene, a dynamic that only becomes truly apparent in the aftermath of automation (or, as Vergès says, in the postcolonial period).

James and Grace Lee Boggs's notion of scientific Black Power is especially useful because they saw automation as a political relationship that attempted to organize racialized and gendered human life as well as nature more broadly for capital accumulation, during a period (as remains true today) in which those connections were rarely made. The Boggses did this not as academics, concerned with making systematic connections, whether between automation and pollution or between Black cities and global anticolonialism, but instead as movement theorists. Thus they were primarily focused on mobilizing this knowledge for the Black freedom struggle in order to create a "new plateau" in human development. As a political project, urban land was necessarily connected to the production of outsiders by automated capitalism and, alongside Indigenous dispossession, to the history of Black struggles for sovereignty and self-determination dating to the post–Civil War Reconstruction demand for land as reparation for plantation slavery.[63] They also shared this concern with other Detroit-based Black radical organizations in the 1960s. For example, in 1969, at the Black Economic Development Conference, former members of the LRBW called for the organization of a "Southern Land Bank" to help set up cooperative farms. During the same period, the founders of the Republic of New Afrika moved from Detroit to rural Mississippi in order to set up a land-based Black-majority independent country.[64]

The Boggses, however, did not return to the land. Instead, their understanding of land and of potential economies that could be created in the aftermath of capitalist automation was rooted in the specificities of Black cities as spaces damaged first by their industrial use and then as disposal sites for waste and pollution, like used "fancy wrappers," "throwaway bottles," and runoff from things like "pesticides" and "enzymes detergents."[65] In this respect, their sense of urban land parallels Anna Lowenhaupt Tsing's notion of a "third nature," or "what manages to live despite capitalism," a concept she names after William Cronon's "second nature," or the transformation of the environment by capitalism.[66] The Boggses, unlike Tsing's work on salvage capitalism in the forests of Oregon and in international matsutake markets, only ever hint at life beyond the human, and their sense of Black Power social practices are far more anticapitalist and oppositional than those examined by Tsing. However, they share with Tsing a sense of urban land as third nature in the aftermath of modernizing, progress-oriented factory capitalism modeled on the plantation. Most important, Tsing and the Boggses both look to practices of life making and conceptions of freedom in these damaged places. These forms of life are full of possibility but are also contradictory. Remember that James and Grace Lee Boggs argued that automation made (especially working-class Black) Americans "expendable to the economy," but it also "liberated" them for the first time since their arrival in the Americas "from the necessity to work on behalf of white development."[67] Urban land is thus an end point in the sense of migrations motivated by multiple—and to some extent contradictory—forms of racism and dispossession and in terms of the toxicity, waste, and health problems that are the aftermath of automated industrial capitalism. However, for scientific Black Power, urban land is also the starting point for political experiments in what life and socially necessary activity can do in third nature.

Dialectical Humanism

The automated geographic reorganization of capitalism that produced, initially and primarily, Black outsiders on urban land also produced a more general, but also specific, version of the human. Access to affluent society was unequally distributed, but Keynesian capitalism in the United States sought to turn Americans into consumers in order to secure effective demand and social stability. The Boggses describe this as a throwaway society dependent on hierarchically racialized labor. In *The American Revolution,* James described this even more forcefully:

> Inside each American, from top to bottom, in various degrees, has been accumulated all the corruption of a class society that has achieved its magnificent technological progress first and always by exploiting the Negro race, and then by exploiting the immigrants of all races. At the same time the class society has constantly encouraged the exploited to attempt to rise out of their class and themselves become exploiters of other groupings and finally of their own people. The struggle to rid themselves and each other of this accumulated corruption is going to be more painful and violent than any struggles over purely economic grievances have been or are likely to be.[68]

Racialized exploitation through technocapitalism dehumanizes "each American" differentially, in large part because attempts to escape exploitation require continued differential exploitation. The starting place for the struggle against this individual and systematic corruption was in the city, where Black outsiders would be guided by the vanguard party. In place of capitalism, scientific Black Power would introduce a sweeping reorganization of institutions. Throughout the cybercultural era (and until their deaths), the Boggses argued that revolution would not be instantaneous but instead a long process that would involve new forms of thought and a variety of practices, like shifting the focus of life from the commodity form labor power to new forms of socially necessary activity in order to create new values. Beginning in 1963, they described this philosophy as dialectical humanism because the American Revolution would be the first to happen in a place of abundance in which human needs could be easily met. This revolution would not be about material production; instead, as Grace Lee wrote in *Correspondence,* it would be "a radical change in man's image of himself and of his rights and responsibilities . . . [and] new revolutionary struggle for new human values and new human relations."[69]

Reflecting on dialectical humanism in *The Next American Revolution,* Grace Lee emphasized that the affluence of the cybercultural era was premised on multiple forms of exploitation and that a classless society could not create a truly postscarcity world if it continued to operate under the same growth- and consumption-based premises as both U.S. and Soviet forms of state capitalism. The struggles of the nineteenth and early twentieth centuries, she argued, had been "not only to correct injustices but also to achieve rapid economic growth." However, in the United States, a country that had "achieved its rapid economic growth and prosperity at the expense of African Americans, Native Americans, other people of color, and peoples all over the world, our priority had to be in correcting the injustices and backwardness of our relationships with one another, with other countries, and with the Earth. In other words, our revolution had to be for the purpose of accelerating our evolution to a higher plateau of Humanity. That's why we called our philosophy 'dialectical humanism' as contrasted with the 'dialectical materialism' of Marxist–Leninists."[70] Following this critique of both outmoded forms of revolution and outmoded forms of the dialectic, Grace Lee Boggs then turned back to *Revolution and Evolution in the Twentieth Century,* where the Boggses had argued that the United States was a country of abundance and that its political economy had too long been dominated by assumptions about scarcity. The new revolution in the United States must not proceed from premises modeled on the exploitation of "scarce" nature and land. Instead, "it is for total political power to make decisions as to what should be done and what should not be done with land. . . . The revolution to be made in the United States will be the first revolution in history to require the masses to make material sacrifices rather than to acquire more material things. We must give up many of the things which this country has enjoyed at the expense of damning over one-third of the world into a state of underdevelopment, ignorance, disease, and early death."[71] As the Boggses note later in *Revolution and Evolution,* technological fixes—because they largely rely on an intensification of extraction, production, and consumption—can no longer be the solution to social and environmental problems; instead, the question is, "Who is going to give up what?"[72]

Bill Mullen notes that dialectical humanism draws on multiple sources—Grace Lee Boggs's discovery and translation in the 1940s of Marx's early writings; and C. L. R. James's and Mao Zedong's work—but it was especially informed by the era's "Bandung humanism," which sought to foster independence and international solidarity and also a broader decolonization of life and thought.[73] In these passages, Grace Lee Boggs also emphasizes that dialectical humanism is a political ecology in the sense that antiracist, internationalist, and class-based projects can no longer be premised on the accumulation models of the past. Instead, these projects must confront questions of how to organize an economy and distribute surplus such that people can reencounter themselves and their lives in ways that will allow them to "relate to nature again."[74] This type of language is often dismissed as too utopian. However, for the Boggses, new relations to nature, whether human or nonhuman, were a political project attentive to the interlinking of GDP growth, organized scarcity, racial inequality, imperialism, and environmental destruction. In these respects, dialectical humanism and scientific Black Power were degrowth projects.

As with political ecology, and to a lesser extent automation theory, the Boggses have virtually never been associated with the degrowth movement—this despite the fact that their work coincides with the earliest antigrowth critiques, and most importantly because their political praxis offers a crucial perspective on the problem of how to transition to a postscarcity classless society not premised on growth. They critiqued growth-based accumulation strategies as early as *The American Revolution,* arguing there that a "social revolution in the United States" must mean production should only be for "social use" and for those "who need it" instead of for always already obsolete consumer objects.[75] This perspective is rooted in a Marxist tradition of worker control and Marx's famous injunction that communism would be a system best described as "from each according to his ability, to each according to his needs." However, their critique of growth and consumerism is not transhistorical; instead, it dates to the moment of what Matthias Schmelzer has called the "hegemony of growth": the post–World War II creation of "national economies" and the calculation of gross domestic product as the sole criterion for describing prosperity.[76] The critiques of the GDP growth model have been numerous, but they became particularly prominent at the time of the publication of *The Limits to Growth* in 1972, and it was in that year that André Gorz coined the term *décroissance.*

Defined simply, degrowth is "an equitable downscaling of production and consumption that increases human well-being and enhances ecological conditions."[77] The degrowth movement is often critiqued as quixotic, and there is increasingly a great deal of debate external and internal to the movement, which intersects with long-standing Marxist discussions about how to transition to a postcapitalist economy not premised on growth.[78] The Boggses' sense of scientific Black Power that organized working-class and dispossessed peoples while attempting to practice new forms of socially necessary activity offers a valuable approach to how a degrowth transition might take place.

During the cybercultural era, through the Organization for Black Power (and later the National Organization for an American Revolution and the Boggs Center to Nurture Community Leadership), the Boggses argued that Black Power could radically alter the form of the city and could create a new economy. In their "Manifesto for a Black Revolutionary Party," they laid out their platform, describing a world in which Black Power would take control of all city institutions and do things like reduce the number of hours people work and eliminate "living space for cars" in favor of "living space for people."[79] In the years to come, they would add to this list the right to a pollution-free environment and the free distribution of birth control, among other things.[80] More broadly, scientific Black Power sought to create a new economy that would deemphasize, and ultimately eliminate, the commodity form as well as the ways that reproduction is devalued and hidden in capitalist economies in favor of a new emphasis on activities like political and community organizing, social services, health care, and education. They put forth these proposals at a moment during which the nature of Fordist work was being broadly called into question and a new emphasis on service industries, information technologies, and increasing participation in the labor market by women was on the verge of radically changing the composition of work in the United States—phenomena that the Boggses in many respects anticipated, and to which they hoped to provide political alternatives.[81] The role of automation was especially important for their conception of work. They argued, "Because labor is becoming more and more socially unnecessary in the United States and another form of socially necessary activity must be put in its place that a revolution is the only solution. And it is because Afro Americans are the ones who have been made most expendable by the technological revolution that the revolution must be a black revolution."[82]

This passage makes explicit the dialectic reason why Black Power must necessarily be at the forefront of the creation of a postscarcity, classless society. It also brings to the fore the significance of forms of "socially necessary activity." Activity was an attempt to both rethink the categories through which they analyzed what people do in capitalism as well as a way to reimagine the nature of class and what counts as participation in the new society. In socially necessary activity, there are echoes of Hannah Arendt's then-contemporary discussions of the *vita activa* in *The Human Condition.*[83] Arendt's *vita activa* is divided into labor (biological processes), work (the making of the world of things), and action (political interaction between people that speak to the plurality of humanity). Like the Boggses, Arendt was rethinking these old categories at a moment when automation and space exploration seemed to be altering how the practices of labor, work, and action conditioned the human. Grace Lee Boggs and Arendt, in fact, briefly debated the politics of the cybercultural era during the Evolving Society conference.[84] The Boggses largely collapse this tripartite scheme into the single Marxist-influenced category of socially necessary activity (as opposed to socially necessary labor time in capitalist production), and they write from within a Black Power movement seeking to revolutionize the organization of society. Yet like Arendt, it is through socially necessary activities, what people do (or, as the Italian autonomists would say, through forms of self-valorization), that a new organization of the human can emerge. They envisioned socially necessary activities as new forms of social reproduction for a classes society but also as the revolutionary ethical-political modalities through which people can become self-conscious of their own life-making capacities beyond a consumer capitalism that is based on racial and gendered exploitation and the appropriation of nature: "in the process of discovering that plenty is available to everyone without the need to grab and hoard and at the same time be involved in making the decisions as to what should be produced and how the community should be organized, people will soon recognize that mere possession is not the height of human value."[85] It is thus especially through forms of socially necessary activity, or new forms of social reproduction, practiced by politically organized outsiders and those in solidarity with them, that degrowth, as part of the long road to a classless society, can be achieved.[86]

As all of this suggests, scientific Black Power's political struggle for a classless society was not utopian in Engels's sense, but instead it learned from previous revolutionary movements and was grounded in the already existing struggles of working and dispossessed peoples at a specific, cybercultural conjuncture in U.S. and planetary capitalism. They saw the citadel of U.S. capitalism in a dialectical way; the automated production of outsiders simultaneously offered new possibilities to reorganize social reproduction and to create new values. They thus drew on already existing practices; like J. K. Gibson-Graham, they emphasized the redistribution of surplus as an "ethical praxis of being-in-common" and that "when appropriated surplus labor is distributed—in the form of payment, gift, or other allocation—community is produced or at least potentiate. . . . Each allocation allows for a becoming, a new way of being."[87] A true cybercultural revolution would thus utilize automation and cybernation technologies to free people from work, but they were not necessary conditions for freedom. In fact, a postscarcity, classless society would reconceptualize freedom in the (corrupt) U.S. sense, defined as the ability to exploit others. It is important here to reemphasize that dialectical humanism, and the Boggses' political ecology more generally, was a project to remake human subjectivities and to expand conceptions of freedom. In *Revolution and Evolution,* they argued that "freedom is not a thing"; freedom is a "relationship" with others that "comes with recognition of the dialectical inter-relationship between oneself and one's social and physical environment."[88] Likewise, the Boggses continually emphasized that ethics of, and claims to, the human are necessarily relational. As with demands for a degrowth politics of giving up, to embrace full humanity is to do so in terms of a responsibility to others, both human and nonhuman. This is to say that dialectical humanism, which happens through the praxis of a community coming into being by justly organizing its surplus, must also necessarily remake people's subjectivities.

To remake desires and subjectivities, James and Grace Lee Boggs called on people to follow "the socialist road" and the "struggle to develop ourselves as new human beings" that was attentive to the histories of racialized dispossession on which (previous) economic growth models and normative figures of the human are premised.[89] As they emphasize in *Revolution and Evolution in the Twentieth Century,* they drew on the knowledge of previous revolutions, especially the French, Soviet, Chinese, and Vietnamese ones, in order to formulate how Black Power would resolve the specific problems of the United States in the late twentieth century. As they were for Sylvia Wynter, who was beginning to develop her theories of "Man" at the same time, the histories of slavery and its afterlives were also of supreme importance to the American and anticolonial revolutions. Wynter, discussing what Black education should look like in the early twenty-first century, argued that the mode of production didn't just produce "material conditions" but also "our conception of being. Every mode of production is a function of producing that conception."[90] She thus asks "what system of economics would we need" in order to secure the well-being of all instead of the current world of abundance that functions only to secure the well-being of the genre of the human she calls "Man" while simultaneously damaging the biosphere.[91] For the Boggses, the struggle to replace the racialized and gendered *homo oeconomicus* genre of the human with a new set of values attendant to the histories of Black and anticolonial struggles for freedom, as they were manifested in the specificities of the cybercultural era, involved new participatory and self-conscious forms of activity on urban land. However, for them, this would be a never-ending process, and one without, as Stuart Hall has said, guarantees.

We can read their most famous statement in this vein: "What time is it on the clock of the world? Is it just 8:30 p.m. on the evening of July 4, 1973? What is time? Is it just the measurement of the distance traveled by the hands of a gigantic clock? How do we estimate time in terms of the evolution and development of humanity?"[92] Nicholas Mirzoeff has said that the "clock of the world" is a clock of revolutionary time: "It is a relation of human and nonhuman life, considered in relation to planetary time. This time is nonlinear and open, offering a means to rethink our relation to the world."[93] In *Revolution and Evolution in the Twentieth Century,* the accelerated time of capital accumulation is posed against, as Mirzoeff suggests, an alternative time of life. As the Boggses wrote, definitions of "growth" must go beyond "quantitative" ones: "whenever you are dealing with living organisms, and especially with human beings, the most important thing is to see them in terms of their development, which can only be evaluated over a period of time."[94] Thus, for humanity to initiate a new plateau, new conceptions of time, value, and how to inhabit the earth as relational beings must be made. In the cybercultural era, this necessarily meant engagement with the specific histories of technocapitalism and the struggles of working and dispossessed peoples. If the cybercultural revolution, as they defined it in "The City Is the Black Man's Land," was a rapidly emerging automated reorganization of race, class, and labor that seemed to be primarily leading to new forms of racialized dispossession and increasing planetary damage, then a countercybercultural revolution could not simply take up capitalist technologies but would necessarily require a new political economy that reoriented the activities that people engage in in order to make a classless society and new human values.

3

Value in the Materialist World

READING NOAH PURIFOY'S *66 SIGNS OF NEON*

Today, the artist Noah Purifoy is probably best known for his environmental sculptures at the Noah Purifoy Desert Art Museum just outside Joshua Tree, California. Here, on ten acres of land, he created more than 120 junk assemblages between 1987 and 2004, the year of his death.[1] Emblematic of this work is *Bessemer Steel,* which looks like a decommissioned factory and is constructed from rusting sheet metal, different types of pipes, an American model 57CR instrument sterilizer, and a rusted gas heater, among other objects. The title has multiple historical and personal references: "Bessemer" refers to Henry Bessemer, who in the mid-nineteenth century invented a fast, cheap, and efficient method of producing steel by using oxygen to remove impurities from pig iron and thus played a significant role in the development of industrial capitalism. *Bessemer Steel* also has personal resonances for Purifoy: he grew up during the Jim Crow era near the industrial town of Bessemer, Alabama (named for Henry), and many of his family members worked in its factories.[2]

Bessemer Steel poses the question of how the rusted junk materials and their reuse critique what Robert Smithson called in "A Sedimentation of the Mind: Earth Projects" the "hard, tough," "permanent," "technological ideology" of steel and are a sign for the often suppressed histories of racialized production and value extraction that technocapitalism has always relied on.[3] This orientation dates to the post–Watts Rebellion exhibition *66 Signs of Neon,* the focus of this chapter. It was at this moment that Purifoy began to construct his sculptures entirely out of recycled junk materials, and when he began to overtly theorize his ethics of poverty and how making junk art was about the process of revaluating not just the things discarded or used up by technocapitalism but also, and especially, the racialized people and communities that have been subject to, as James Boggs wrote, the "wear and tear of constant use" by that system of (re)production.

The question of value as it is manifested in Purifoy's *66 Signs of Neon* exhibition and the way he is in dialogue with James and Grace Lee Boggs animates this chapter—a shared ethical perspective perhaps influenced by the fact that James Boggs attended high school in Bessemer.[4] Black Power responded to the cybercultural conjuncture in multiple theoretical and political ways, such as reconsidering the political geographies of imperialism and the nation-state, the role of the welfare state and state capitalism, and how forms of production and reproduction were being reorganized, as well as historicizing racialized technological change. These issues, although not synonymous, are all connected, and all are implicitly or explicitly about value—both how it is organized by capitalism and what alternative forms of value might emerge through anticapitalist praxis. Noah Purifoy, whom Kellie Jones has referred to as the "Philosopher-Artist," is rarely, if ever, described as a theorist of technocapitalism.[5] Here, to extend chapter 2's concerns while approaching them anew, I argue that Purifoy's theoretical and artistic work in fact provides an incisive perspective on how cybercultural capitalism organized racialized hierarchies of life in relation to longer geologic and African diasporic histories, and that his post–Watts Rebellion aesthetic practice provides a model for the types of activity out of which new values can emerge.

Before becoming one of the key figures in the Los Angeles assemblage art movement, Purifoy was a social worker, one of the first Black students to attend the Chouinard Art Institute (now CalArts), and a furniture maker and interior designer. He was also the first director of the Watts Towers Arts Center and a founding member of the California Arts Council. In this chapter, I primarily focus on Purifoy's largely unexamined essay, "The Art of Communication as a Creative Act," written with Ted Michel, that appeared in the *66 Signs of Neon* exhibition catalog. In a sometimes oblique, impressionistic, and provocatively abstract four-page essay, he echoed the Boggses' discussion of outsiders and capitalism: "Watts finds itself virtually set down in the center of junk piled high on all sides. Its main industry is junk! The essential question being posed to the community by the exhibit was, what is the true value of these materials over and above sale to junkyards for a few cents?"[6] This is a question not just about the value of "junk" but also about how Black people in mid-twentieth-century Watts, Los Angeles, were socially differentiated and devalued. In the 1997 text *High Desert,* Purifoy wrote that junk sculpture was "a concept that, by taking junk and making it into something else, can help make people realize that their junk, their lives, can be shaped into something worthwhile."[7] Junk sculpture, which Purifoy understood to be part of a creative process that exceeded in importance the art object and that was applicable to all aspects of life, is thus not just about devaluation but, more important, is about how it can be the basis for new and "creative" forms of value alternative to those of what he called the "materialistic world."

The *66 Signs of Neon* exhibition, initially produced for the 1966 Watts Renaissance of the Arts Festival, was organized by Purifoy, along with Judson Powell, Debbie Brewer, and others, and was a collection of junk assemblage sculptures made from the damaged artifacts and material remnants that they collected in the days after the August 1965 Watts Rebellion. The exhibition traveled around the United States and then ultimately went to Germany in 1972 before returning to the United States and getting junked because the sculptures were in such bad shape after years of being mishandled.[8] Thus, while the sculptures are important, they are now largely lost as physical objects, accessible only through the photos in the *Junk Art: 66 Signs of Neon* exhibition catalog and Purifoy's own "evolving system of philosophy."[9] Most critical work on *66 Signs of Neon* focuses on the immediate links between the exhibition and the Watts Rebellion. These links are undoubtedly central; indeed, Purifoy told Karen Anne Mason in 1990 that the Watts Rebellion "made" him an artist.[10] However, in "The Art of Communication as a Creative Act" (as well as in later interviews and essays), Purifoy emphasizes that the exhibition utilizes "the August 11 event but that it transcends it, and rises above social protest." For Purifoy, the debris from the Watts Rebellion is only one form of junk.[11]

As I argue in this essay, *junk,* as Purifoy uses it, is a more specific term than *waste* or *garbage* or *trash* because it is a concept that opens specific material and political-economic histories. Junk resonates with what Jussi Parikka terms media geology, or a focus on the geologies and material afterlives of technologies, and mid-1960s Watts, Los Angeles, is one of the places that commodities and technologies, media and otherwise, went after they had been used up and thrown away.[12] The Watts "junkers of the community," of which Purifoy was one, earned a living in the informal economies in the shadow of Southern California's high-tech and aerospace industries. Purifoy's junkers are, in this respect, figures of Michael Denning's "wageless life."[13] For Black Power theorists, the cybercultural conjuncture was not simply revolutionary in the technological sense; they also conceptualized it as a specific conjuncture in the ongoing differentiation between waged and unwaged work constitutive of technocapitalism understood as a system conterminous with plantation slavery and its afterlives.

In different ways, for the Boggses, King, and others, the conjuncture was contradictory because it created the conditions through which Black peoples no longer had to necessarily be productive for White wealth. It opened up the possibility of new forms of freedom, but also of new forms of dispossession and expendability. It is precisely this contradiction that made outsiders a potentially revolutionary force. Purifoy's approach to this is less politically revolutionary; however it is also expansive, if grounded, in ways similar to the Boggses' belief that a "new plateau" for humanity was possible. Purifoy's "community" of "junkers" labored to earn a living in this contradictory high-tech world. However, they were also creative laborers—both in quotidian ways and as a community that, through junk assemblage in spaces "outside of normal circumstances," as Purifoy writes, like the Watts Towers Arts Center—who created new forms of value alternative to technocapitalist valuations of life. *Junk* thus signifies reuse and not simply waste, and junk assemblage, as I will discuss, was a means through which imbricated social, economic, and geologic histories of the conjuncture are examined, negotiated, and reworked. As Kellie Jones notes, junk "assemblage allowed one to evade the fixity of blackness—its looks, meanings, materiality, and form."[14]

In the last section of this chapter, I turn to how Purifoy conceptualized reorganizing junk as a creative practice that rethought normative forms of art and productivity. In doing so, I link his approach to C. L. R. James, Grace Lee Boggs, and Cornelius Castoriadis's discussion of worker "self-movement," "creativity," and "universality" in the pamphlet *Facing Reality* and how this encounter opens a way of seeing how junk assemblage's "evasion of the fixity of blackness" is connected to a longer history of African diasporic "creative" resistance to normative forms of productivity and labor. This will be preceded by a discussion of the theory and materiality of junk as Purifoy conceptualized it in "The Art of Communication as a Creative Act." As in chapter 2, here I approach Watts as a center for planetary capitalism and of (via Vergès) the racial Capitalocene. The normative subject of the geologic era most often called the Anthropocene is for Vergès, like Sylvia Wynter, the genre of the human called Man2 or *homo oeconomicus*.[15] In renaming the epoch, Vergès draws attention to the histories of dispossession, racialization, capital accumulation, and waste production that are constitutive of planetary transformation and asks, "What connection can be made between the Western conception of nature as 'cheap' and the global organization of a 'cheap,' racialized, disposable workforce, given the conception of nature as constant capital and the fact that 'the organizers of the capitalist world system appropriated Black labor power as constant capital'"?[16] Reading Purifoy's conceptualization of junk helps us to see the cybercultural revolution as not only a moment of automation and cybernation but also one in which these technologies are part of a larger organization of extraction, production, use, and junk in which capitalist forms of value are produced at every moment of this process, including through the labor done by the junkers of the community.

However, I first turn to the most specific instantiation of the materialistic world cited in "The Art of Communication as a Creative Act": the official California State report on the Watts Rebellion, *Violence in the City—An End or a Beginning?*, better known as the McCone Report, after its lead author, former CIA director John A. McCone. Purifoy notes, "the validity of its [junk's] use as an art expression did not occur to us until we came across the McCone Report. . . . We thought it a good report. . . . However, we observed that there was no mention of art education. Education, yes, but not art education. This was our cue. We decided that *66 Signs* could establish its validity by becoming an addition to the McCone Report."[17] Roderick Ferguson and Daniel Widener argue that the McCone Report was central not only to how Purifoy conceived of his art but also to what Widener called the wider "battle over the production of knowledge" in post–Watts Rebellion culture.[18] As in my discussion of *The Triple Revolution* and *Technology and the American Economy* in chapter 1, I look specifically here at how the McCone Report was a document of cybercultural capitalism that, while ostensibly a report on the Watts Rebellion, defined Watts as part of the history of space-age United States, where mechanization was altering the ways that nature and Black Americans had historically been put to work by and for capitalism. Purifoy's creative reevaluation of junk and people is thus an addition to this understanding of how technocapitalism organized and disorganized life in Watts.

Mechanization, Operational Knowledge, and the McCone Report

"By September," Purifoy writes, "we had collected three tons of charred wood and fire-moulded debris." The question then became what to "do with the junk we had collected, which had begun to haunt our dreams." At the same time as this interstitial moment—specifically August 24, when Purifoy and his associates were "haunted" by junk and planning a sculpture garden at the Watts Towers Arts Center instead of an exhibition—California Governor Edmund Brown created a commission that sought to "probe deeply the immediate and underlying causes of the riots," to understand how to prevent future "recurrence," and "to develop a comprehensive and detailed chronology and description of the disorders."[19] *Violence in the City—An End or a Beginning? A Report* by the Governor's Commission on the Los Angeles Riots was published on December 2, 1965. The Watts Rebellion began on August 11, 1965, with an altercation between police and community members after the arrest of Marquette Frye for an alleged traffic violation. Before it ended, on August 16, thirty-four people would be killed. The McCone Report was produced over three months and was a condensed version of eighteen volumes of research, which, while seemingly extensive, was critiqued as insufficient given the significance of the event.[20] In brief, the McCone Report understood the rebellion in the "curfew area" to be the result of mass unemployment, but it also highlighted the neighborhood's lack of public transportation, lack of respect for the police, and the largely Southern migrant population's excessive "expectations" for a neighborhood that the McCone Report famously called neither a "slum" nor an "urban gem." The solution? Although there was "no immediate remedy," the report recommended a number of "costly" employment and education programs that would "take time." Most important, the community needed to gain a new level of respect for the police, and Black leaders needed to "shoulder a full share of the responsibility" for the "well being" of the community.[21] In "The Art of Communication as a Creative Act," Purifoy called it, with qualifications, a "good report," although decades later, he was very critical of it.[22] It was, however, largely derided by the Watts community.[23] As Daniel Widener writes, the McCone Report rested on a "culture of poverty" thesis similar to that espoused by the Moynihan Report, also issued in 1965, which notoriously blamed Black families for issues of structural poverty.[24] Bayard Rustin was a particularly harsh critic of the report, and in his review in *Commentary,* he blasted the report's central contention that residents who had moved from the South were unprepared for city life.[25]

Edward Soja has been equally critical of similar narratives about Watts. The rebellion was, without question, motivated by racial segregation in housing and the virulent racism of the Los Angeles Police Department, particularly the notorious chief of police, William Parker. Most significant, however, for Soja—much like James and Grace Lee Boggs in "The City Is the Black Man's Land," although with the advantage of historical perspective—the Watts Rebellion was a signal moment in a capital accumulation "crisis" that "marked one of the beginnings of the end of the postwar economic boom and the social contract and Fordist/Keynesian state planning that underpinned its propulsiveness."[26] Soja notes two important features of the Los Angeles economy. The first is that well-paying unionized industrial jobs were available close to Watts, but racist hiring practices prevented residents from accessing those jobs. Second, unlike most stories of deindustrialization, Los Angeles never deindustrialized; instead, the period's crises were part of a shift to post-Fordist, "flexible" industries.

That the Watts Rebellion marked both a moment of crisis and a transition in racial capitalism was not lost on either the McCone Report's critics or, to some extent, the report itself. Bayard Rustin, for example, brought out the extent to which the linear teleology of the

McCone Report tried to establish the rebellion's causal relationships while also attempting to reestablish the liberal capitalist future that it suspected had been altered by the "explosion" and the "formless, quite senseless" violence of the rebellion.[27] As Rustin writes, the report emphasized the "inevitability" of Black "behavior," and by "tacitly assum[ing] that the Civil Rights Acts of 1964 and 1965 [were] already destroying" segregation, it was free to engage in an abstract debate over "data" and the "problems" of housing, education, and employment "without attacking the conditions of de facto segregation that underlay them."[28] According to Rustin, if the McCone Report ostensibly emphasized the rebellion's multiple causes, it in fact reinscribed it within a linear teleology of interracial overcoming formalized by the institution of legal rights on the federal level—an overcoming that had not yet actually come to pass and, according to Rustin, would not, given the McCone Report's assumptions.

Alongside this, the McCone Report is also a technological narrative akin to other institutional documents from the period, like *Technology and the American Economy* and, less explicitly from a technological perspective, the Kerner Commission report. As it says, "Of what shall it avail our nation if we can place a man on the moon but cannot cure the sickness in our cities?"[29] This captures the contradictions of the McCone Report's general conception of history and futurity. Here the space age that is only technically yet to come (in the 1969 moon landing) is uncomfortably imbricated with a pathologized and racialized urban space that does, and does not, call into question the space-age technological future. The McCone Report's ultimate concern is not with the six days of the Watts Rebellion but instead with Black conformity to its space-age temporality and an insistence that the U.S. Cold War narrative of space-age technological progress should not be called into question. If this were to happen, the report says, "costly" education and employment programs (along with Black leadership) must be a training ground for this perhaps impossible reincorporation. Roderick Ferguson argues that the report is a document of visual culture that trains readers to see the Black community as socially dysfunctional.[30] One example and remedy, so to speak, to this is that the report insists that education programs should focus on "'attitudinal training' to help . . . [Black students] develop the necessary motivation, certain basic principles of conduct, and essential communication skills, all of which are necessary for success" in the White world of productive labor.[31]

If Purifoy's aesthetics and "art of communication" were conceived as an "addition to" the McCone Report, then it is in part a critique of this reductive understanding of communication as "attitudinal training" that the report contextualized in relation to Watts residents' supposed excessive "expectations" and the insistence that they were "unprepared" for "modern city life."[32] This history is introduced in the line "idleness brings a harvest of distressing problems."[33] *Harvest* is a provocative word choice. In part, the report's linear trajectory of (un)freedom in the U.S. space age is one where Watts is constituted by the legacies of the plantation and Black migration from the South to Los Angeles.[34] The migration, according to the report, was specifically a history of "mechanization":

> Modern methods and mechanization of the farm have dramatically, and, in some regards, sadly reduced the need for the farm hand. With this, a drift to the city was the inevitable and necessary result. . . . But the conditions of life in the urban north and west were sadly disappointing to the rural newcomer, particularly the Negro. Totally untrained, he was qualified only for jobs calling for the lesser skills and these he secured and held onto with great difficulty. Even the jobs he found in the city soon began to disappear as the mechanization of industry took over, as it has since the war, and wiped out one task after another—the only tasks the untrained Negro was equipped to fill.[35]

This narrative, while diverging politically, parallels the history of the cybercultural revolution as conceived by James and Grace Lee Boggs. For them, automation was only the most recent moment in a history of mechanization and was a racialized and geographic reorganization of labor and capital. This "undevelopment" had rendered a large percentage of the Black population outsiders in cities after having been, as they put it in "The Myth and Irrationality of Black Capitalism," "exposed to the most modern appliances and machinery."[36] Both of these narratives resonate with Purifoy's claim that "Watts finds itself virtually set down in the center of junk piled high on all sides." In Watts, racialized working people and junk are produced in common; they are what remains after Black labor is subject to the "wear and tear of constant use" of U.S. technological modernity.

The McCone Report's conception of mechanization, which it sees as part of a space-age futurity, largely ignores the "wear and tear" of that development, and especially how it is part of a history of not only migration but also of putting human and extrahuman nature to work for capital accumulation. This "attitude," to use the report's own language, is what Herbert Marcuse defined as one-dimensionality and "operational" knowledge, or the "theory and practice of containment," a world in which negative or critical, and thus political, thought had been rendered virtually obsolete.[37] In *One-Dimensional Man* and *Eros and Civilization* (especially its 1966 "Political Preface"), Marcuse charts the midcentury development of what he calls a repressive "unfreedom" through increasing technological rationality. He writes in the introduction to *One-Dimensional Man* that capitalist technological modernity

> is one "project" of realization among others. But once the project has become operative in the basic institutions and relations, it tends to become exclusive and to determine the development of the society as a whole. As a technological universe, advanced industrial society is a political universe, the latest stage in the realization of a specific historical project—namely, the experience, transformation, and organization of nature as the mere stuff of domination. As the project unfolds, it shapes the entire universe of discourse and action, intellectual and material culture. In the medium of technology, culture, politics, and the economy merge into an omnipresent system which swallows up or repulses all alternatives. The productivity and growth potential of this system stabilize the society and contain technical progress within the framework of domination. Technological rationality has become political rationality.[38]

Pamela M. Lee succinctly summarizes Marcuse's arguments on technological rationality thus: "We could put it crudely: postwar technology is organization."[39] Marcuse's work is relevant because the organization and domination of nature as the historic project of the racial Capitalocene are even more apparent today, and because it shares with Black Power cyberculture a sense that automation is *potentially* liberating. Marcuse argues that technologies like automation, which could reduce the need for labor, correspondingly make the need for repression, understood as the psychic obligation to be value productive for the capitalism, obsolete, and thus psychic liberation becomes possible for the first time—at least in capitalist modernity. It is because of this potential that critical thinking and oppositional modes of existence necessarily became incorporated into the system, and an operational thought that was reproductive only for the capitalist ideology of scarcity came to dominate.[40]

The McCone Report is part of this materialistic world not only to the extent that it writes a chronological history of United States cybercultural modernity that leaves Black working people outside of this development, but also to the extent to which it tries to produce a subject appropriate to this form of organization. By undergoing "attitudinal training" and accepting "law enforcement," people may be organized into a modernity of linear progress and operational thought, but they are simultaneously undeveloped for outdated and/or precarious labor. This is a problem that I will return to in chapter 5 on the League of Revolutionary Black Workers and in a different way in the final chapter on the Black Panthers. Of course, the ideal reader of the McCone Report is the White bourgeois subject of Marcuse's text—those taken care of by Keynesian state capitalism. In the 1966 "Political Preface" to *Eros and Civilization*, Marcuse addresses more directly the racialized forms of exclusion and waste production at the heart of the post–World War II space-age stability that the McCone Report desires: "Affluent society depends increasingly on the uninterrupted production and consumption of waste, gadgets, planned obsolescence, and means of destruction, the individuals have to be adapted to these requirements in more than the traditional ways. . . . This freedom and satisfaction are transforming the earth into hell. The inferno is still concentrated in certain faraway places: Vietnam, the Congo, South Africa, and in the ghettos of the 'affluent society': in Mississippi, Alabama, and in Harlem. These infernal places illuminate the whole."[41] In both these lines, and in his argument that the 1960s had, through technological development, become a postscarcity world in which there was a "hierarchical distribution of scarcity," Marcuse begins to move toward a racial capitalist reading of modern technological production that shares much with Black Power cybercultural thought and that specifically approaches Purifoy's understanding of Watts as a community of junkers.[42]

In *One-Dimensional Man* and *Eros and Civilization,* Marcuse tried to produce a critical thought capable of countering the one-dimensionality of a technologically rationalized consumer-warfare state that is based on the domination of nature. Purifoy's aesthetic also took up this task and supplemented the educational project of the McCone Report with one that contested the ways that racialized subjects and junk were distributed in common, in favor of noncapital productive forms of creativity that sought to realize human potential. In the *66 Signs of Neon* exhibition, the emblematic sculptural figure of both the Watts Rebellion and the damage of hierarchical distributions of scarcity is the now-lost work *Sir Watts,* which was assembled out of the remnants of the Watts Rebellion to look like medieval armor with a torso of metal; metal mesh where a face is supposed to be; an open chest constructed of pins, silverware, and other similar items covered by a piece of clear plastic; and a belly of wooden drawers. Kellie Jones notes that *Sir Watts* signifies both "people in battle" and "a figure simply in want of love and care."[43] Alongside this, following the Boggses and Marcuse, we can also see *Sir Watts* as a figure subject to the "wear and tear of constant use" in a racial capitalism of undevelopment and organized scarcity. However, it is also a figure in this infernal world that, although made of the junk gathered in the informal economy at the edges of the aerospace industry, illuminates through its construction those social relations, and in doing so, it begins to open the possibility of, as Purifoy says, human potential and a creative reevaluation of those social relations.

A Junk Aesthetic for the Cybercultural Era

In his 1990 interview with Karen Anne Mason, Purifoy said that during the 1960s, while he "could have been not poor . . . I wanted to experience the ins and outs of poverty so that I could report it as it was. Which accounts for the media used to express my feelings about aesthetics."[44] Purifoy's "addition to" the McCone Report should be read as a "report" on the cybercultural organization of scarcity, poverty, and junk. As he wrote in "The Art of Communication as a Creative Act," "human potential" and "education which stimulates the whole process of learning" had to be worked out through the so-called impoverished materials immanent to the community. It is in this sense that *66 Signs of Neon*'s junk was "a utilization of the August 11th event, but that it transcends it." Purifoy consistently emphasizes the importance of understanding creativity, and assemblage as reuse and reorganization, as a "becoming": "If junk art in general, and *66* in particular, enable us only to see and love the many simple things which previously escaped the eye, then we miss the point. For we here experience mere sensation, leaving us in time precisely where we were, being but not becoming."[45] This aesthetic of transformation is a sociopolitical becoming that reflects on the temporalities and histories of the relationship between Watts and junk, where junk aesthetics exceed the rebellion and point not only to broader technopolitics of the cybercultural conjuncture but also to the historical trajectories of the racial Capitalocene. Immediately after his reflection on relations of value in Watts, quoted at the beginning of this chapter, he writes: "On another level, the assemblage of junk illustrated for the artist the imposition of order on disorder, the creation of beauty from ugliness. Its analog was the essence of communication, for the placing of unrelated objects in a pattern conceived by intellect and emotion made them speak coherently, much as the juxtaposition of two persons in a certain context permits them to converse in a way that would be impossible under so called normal circumstances."[46] This gives a distilled sense of the ethics of Purifoy's aesthetics, whose concern with communication is not simply an addition to the McCone Report but a fundamental critique of its materialistic world. In this passage there is a tension between "order" and "disorder." Watts is, on the surface, a space of disorder; however, his discussion of the value of junk and racialized human life demonstrates the ways that Purifoy understood this to be an organized disorder. The task of the assemblage artist working in and with the junkers of the community is to make connections that bring to the fore the histories of junk production in the shadows of the high-tech space age and out of which new practices of value emerge.

Purifoy's assemblage aesthetic of new juxtapositions outside of the "normal circumstances" of disorder is visually rendered throughout the *Junk Art: 66 Signs of Neon* catalog, whether in works by other artists, like Arthur Secunda's dense accumulation of circular metallic objects, *The City;* group efforts that comment on the segregation of public facilities and racialized attributions of cleanliness and uncleanliness, like *The Sink;* or in Purifoy's own work. The dialectical relationship between order and disorder is explicitly rendered on the catalog's cover. Against an orange-colored background photograph of a junkyard, the words "Junk Art" hover in large bold black lettering along the top, with the exhibition title in smaller print at the bottom. The title and the orange junkyard are juxtaposed with a disproportionately enlarged and superimposed photograph of an abstract sculpture on the cover's center, entitled *Breath of Fresh Air.* The work is a pedestal made of a base that has attached to it two pieces of relatively smooth metal, one a stovepipe and the other a concave form, atop of which is fixed a third heavily twisted, crumpled, and dented piece of metal that evokes the ejected remains of a car crash or something torn away from a larger structure. Ferguson has described this work as elevating Watts, after the smoke of the rebellion has cleared, to a "new status" of "dignity" counter to the McCone Report's rendering of it as a doomed place.[47] While *Breath of Fresh Air* itself is attributed to Purifoy, as it appears on the cover of the *66 Signs* catalog it is also a collaborative work between Purifoy; the photographer, Harry Drinkwater, who is responsible for the photograph of the junkyard; and Al Khanasa, who did the layout and design. In this sense, it should be read here as an example of junk assemblage philosophy that as a collective work is a transvaluation of impoverished materials in which the sculpture, as a superimposition, is set off from the junkyard in a space outside of "normal circumstances," but which, as a dialectical relationship between the two, creates new forms of order.

The front cover offers a visual entry point into the philosophy and social concerns of the essay "The Art of Communication as a Creative Act." Purifoy's Watts of junk "piled high on all sides" that we glimpse in the orange-hued *Junk Art: 66 Signs of Neon* cover is thoroughly (dis)ordered by race and class. In 1990, Purifoy said:

> In Watts it [junk] was extremely accessible . . . it relates to poor people. Wherever there are poor people, there's piles of junk. People bring the junk there. In Watts, there were mounds of scrap metal all over the place and defunct foundries where there were piles of metal and junk. Garbage day was a time when people put their trash out, but it was often not picked up, and so it stayed there for weeks. . . . So these are characteristics of poor

> people which made it a haven for me—one who collects objects. There's something else about objects that appeals to me—it stimulates my imagination. I can think to do something with it, turn it into something else other than what it was originally designed for.[48]

Purifoy seems to naturalize the "characteristics of poor people." However, this thicker description of Watts also brings out the historicity of junk, which takes us to the motivation behind his ethics of poverty as well as to the ethics of becoming. Watts is a community of junk because of the way mid-twentieth-century U.S. capitalism distributed hierarchies of people, industrial detritus, and toxicity in the so-called inner city. However, while the characteristics of this distribution are specific to the cybercultural era, the proximity of hierarchically racialized and impoverished people and junk objects is central to the technocapitalist organization of life and space more generally. Purifoy's Watts appears peripheral, as people and objects differentiated as junk at the edges of the capitalist world system. However, read through Françoise Vergès's argument that the exploitation and production of race is central not just to the extraction of value but also to its material afterlives, and through Black Power cyberculture's sense of the conjuncture as a transitional moment in technocapitalism, Watts is better seen as one of the centers of this system. Its residents are socially differentiated and used in surplus-value production, forced to make a living in informal economies in the shadow of the aerospace industry at a moment in which there is a shift to flexible forms of labor and strategies of accumulation that are simultaneously new and consistent with long-standing modes of organization.

For Purifoy, although junk may be produced by either the history of racial capitalism or the Watts Rebellion, junk inhabits a different temporality, although one dialectically linked to temporalities of automation and cybernation conceptualized as a rapidly transforming mode of production by James and Grace Lee Boggs, Martin Luther King Jr., or alternatively the McCone Report. Junk is at least momentarily inert, an extended present of "defunct foundries" "haunting" his dreams, "piled up on all sides," awaiting activation. However, Purifoy's aesthetic does not wish to merely reflect, as he writes, this "experience of mere sensation," whether this be the sensation of junk as that which is left over after use or of technopolitical acceleration. Instead, he seeks to facilitate a becoming and to assemble new "contexts impossible under so-called normal circumstances." This is how Purifoy understands creativity as a new space of communication at the boundaries of a "normal" procession of linear time that demands that Black residents of Watts reorder themselves into redundant subjects of U.S. space-age technological progress.[49] The question of the becoming of junk takes us to the heart of Purifoy's aesthetics as an interweaving of temporalities and histories: collecting and selling junk, transforming it through art practice, and making connections between these practices in the cybercultural era. How are they part of historic Black struggles to create new values alternative to the logics of racial capitalism? I turn to this question next. Significantly for thinking about the longer histories of the technologies of the cybercultural era, the materiality of junk also takes one to the geologic times of those technologies, represented most explicitly in the title of the exhibition itself: *Signs of Neon.*

Critical work, whether with regard to *66 Signs of Neon* or Purifoy's practice more generally, tends to focus on assemblage sculptures that involve the suturing of objects.[50] This is undoubtedly Purifoy's primary practice. However, the title of the exhibition is drawn from a different type of junk sculpture, as Purifoy said in 1990: "*Signs of Neon* is a little confusing unless one is aware of where the exhibit got its title from. The exhibit got its title from the drippings of neon signs upon the ground, formulating crystal-like meltings, mixing with the sand and dirt, formulating extremely odd shapes. What Judson and I did was simply take the shapes out of the sand, brush them off, mount them on something, and sell them. They went like hotcakes."[51] This suggests how prevalent neon signs were in Watts at the time of the rebellion. Yael Lipschutz, in "*66 Signs of Neon* and the Transformative Art of Noah Purifoy," reads the neon pieces as critiques of the commodity. The neon sculptures, and the exhibition more generally, embody the arbitrariness of commodities as, following Karl Marx, only signs of exchange value and not of the thing itself. She writes, "In their liquefied, dissolved, disintegrated forms, then, Purifoy and Powell's lead drippings not only express, but also embody the arbitrariness of the commodity-fetish. In so doing, they allude to the object's other, social, even festive, potentials. (After all, neon is the light of the night, and the majority of the signs on 'Charcoal Alley' [a section of 103rd Street] were advertising liquor stores)."[52] Here, *66 Signs of Neon* traverses a transformational temporality: from advertising for other commodities, the signs are transformed into the signs of the rebellion, in all of its multivalent festive and violent modalities, only to then be transformed, through art, into new commodities. *66 Signs of Neon* and this trajectory point to the complexity of commodities as they connect with race and art practices at this moment. But in addition to the history of commodities, neon—and ultimately its transformation—should also be read as embedded in a broader materialistic world of the period, especially in relation to Marshall McLuhan's discussion of light and technological extensions.

For McLuhan, famously, media technologies are important not because of their content but because of their mediums and the ways that they alter the quality and the "scale" of human relations: "the personal and social consequences of any medium—that is, of any extension of ourselves—result from the new scale that is introduced into our affairs by each extension of ourselves, or by any new technology."[53] In "The Medium Is the Message," "electric light" is an especially prominent example of the ways that "the content of any medium is always another medium."[54] "The electric light escapes attention as a communication medium just because it has no 'content.' And this makes it an invaluable instance of how people fail to study media at all. For it is not till the electric light is used to spell out some brand name that it is noticed as a medium. Then it is not the light but the 'content' (or what is really another medium) that is noticed. The message of the electric light is like the message of electric power in industry, totally radical, pervasive, and decentralized."[55] Reading *66 Signs of Neon* and "The Art of Communication as a Creative Act" through McLuhan challenges us to think about the multiple valences of Purifoy's aesthetic thought and practice. Neon signs themselves are uprooted and signify the (dis)order of consumer capitalism and the (dis)order of the wear and tear of modernization. But they also point to the ways that the message of *66 Signs of Neon* is not the value of art objects. Instead, junk assemblages and creative acts are a medium for a new arrangement of social relations beyond the infrastructures of capitalism.

However, neon signs as "lead drippings" open up another question related to the geology of the racial Capitalocene. What happens to light, understood not just as a medium but also as a technological assemblage, when heated or put under pressure and when one of its mediums, in this case neon, escapes and the assemblage is transformed into the lead drippings that were its other medium? Neon and light cease to be either that which compresses space and time or that which extends social/advertising relations, and in their new form, they become (or return to) something more akin to extensions of geologic time (assembled with its continued exchange value as a commodity in the art marketplace). Smithson was critical of McLuhan both because of his anthropocentrism and because he undertheorized technology's materiality. "The manifestations," Smithson writes, "of technology are at times less 'extensions' of man (Marshall McLuhan's anthropomorphism), than they are aggregates of elements. Even the most advanced tools and machines are made of the raw matter of the earth. . . . The tools of technology become a part of the Earth's geology as they sink back into their original state. Machines like dinosaurs must return to dust or rust."[56]

Smithson's essay draws connections between the technogeologic and what he considers the geologic nature of thought itself, as well as McLuhan's theory of technology and the values of art criticism. But it is especially concerned with time. Technologies are part of longer geologic processes that point both to the past (mining) and the future (rust and junk), as well as to the time of the artist who engages with these temporal scales, and who is thus at odds with surplus-value extractive capitalism. "Technological ideology," he writes, "has no sense of time other than its immediate 'supply and demand,' and its laboratories function as blinders to the rest of the world."[57] This relationship extends into the market-driven art world, where art is simply valued as a commodity and thus inversely (under)values the "time" of the artist.

Smithson's essay is a world of market alienation and technological ideology that demands that the artist conform to its time and value structures. In the essay, the artist is presented as a heroic figure who tries to reassert a relationship to the longer temporalities of geology in particular. Smithson and Purifoy's thought are complementary; however, *66 Signs of Neon* is more social, and thus political, and less focused on the individual artist. Purifoy's Watts is also an alienated space of racialized and classed technological ideology and commodity production. Throughout, *66 Signs of Neon* and "The Art of Communication as a Creative Act" problematize this materialistic culture by questioning the way that racialized and impoverished people and industrial junk are valued. The way that neon is mobilized takes us most directly (if also indirectly) to the assembled, arranged, and historically contingent space of Purifoy's aesthetics. Neon is a sign of Watts as a space where racialized subjects are surrounded by junk "piled high on all sides," by the signs of a prior commodity form that was largely unaffordable to the community. The Watts Rebellion contested this social logic, and Purifoy's art, as he writes toward the end of "The Art of Communication as a Creative Act," also sought to "effect" a lasting "change."[58]

Here change is effected in different ways. Through neon, it takes us to the histories of the object and their material conditions of possibility, both geologic and in terms of the racialized forms of labor power and social differentiation that are part of the process of transforming the earth's materials into commodities—and ultimately into junk. Attending to the value of junk and people doesn't take one to a McCone-style educational project of attitudinal training for outmoded labor in the cybercultural era; instead, it opens up temporal and historical scales. The junk itself takes us to both the geologic time of lead and also to the times of mid-twentieth-century technocapitalism, where objects and racialized people are differentiated as the impoverished remains of the logic of surplus-value extraction. Purifoy's junk aesthetic is in this respect essential for conceptualizing the conjuncture as one in which the racialized reorganization of labor and capital is also part of longer trajectories of capitalism's racialized hierarchies of extraction, production, consumption, and disposal.

This totality is evidenced in *66 Signs of Neon*'s earthly materiality. If the sculptures are signs of the "arbitrariness" of the commodity, then they also embody, as mounted lead, the materiality of their geologic origins. Kathryn Yusoff argues that the Anthropocene and the field of geology are best defined as an "atemporal materiality" that she calls "White Geology," "a mythology of disassociation in the formation of matter independent of its language of description and the historical constitution of its social relations."[59] "White Geology" imagines the Anthropocene as a nonracialized geologic epoch where plantation slavery and other forms of racialized labor, like mining, which are in fact constitutive of the epoch, are erased. As sculptures assembled by Black, community-based artists, *Signs of Neon* foregrounds not an atemporality but instead the labor of transformation and the multiple material histories of the object. In McLuhan read through Smithson, light is not only a technological medium but also signifies a technological organization of people, materials, and modes of production. Purifoy, along with Judson Powell, takes the material of the world—the lead drippings of neon signs and other forms of junk—and, by reorganizing dominant relations, transforms them into new ones. But as we saw earlier, this is not simply about art; it is also about an aesthetic practice that rearranges social space, one that puts people in new "contexts" that allow them to communicate "in a way that would be impossible under so-called normal circumstances." Purifoy's assemblage art of communication can, through *66 Signs of Neon,* be seen as a critique of McLuhan as well as a critique more generally of the technological forms of communication and normal circumstances that are not just extensions of people but of hierarchical social relations. What is needed for new forms of communication are new practices of creativity. Assemblage thus not only mobilizes material histories but also produces the conditions of possibility for new horizons of the future and new values; in the Boggses' language, it will initiate a new plateau. This form of creativity, located in the immediate context of the mid-twentieth-century United States and one that rethinks the assumptions of the McCone Report, ultimately articulates with African diasporic resistance to capitalist logics of race and productivity.

Creative Universality

"The Art of Communication as a Creative Act" ends with a speculative consideration of both the present conditions of "every human" in the "materialistic world" and the necessary link between "communication" and "creativity." "Communication" and "creativity," like "junk" and "neon," are located in dense webs of meaning that transform the technocapitalist connotations of those terms and articulate them with a longer history of African diasporic practices. "The philosophy of *66 Signs* postulates that the activities of each and every human being are attempts to dispel his peculiar anxieties. If the outlet chosen is creative, then the nature of individual experiences change, evolves with this activity. If the outlet is materialistic, then the dispelling of anxiety is purely transient, and without permanent benefit . . . [because of shared anxiety] we all are brothers, and the possibility of mutual understanding, as well as communication, exists."[60] Kellie Jones argues that Purifoy was broadly engaged with existentialism in the 1960s, and his emphasis on human "anxiety," with its Kierkegaardian and Sartrean implications, suggests this.[61] While the language of anxiety was popular at the time, Purifoy conceptualizes anxiety's relief in terms of two specific modes: the materialistic and the creative, with the former historically specific and false and the latter universal and true. Creativity is always linked here to establishing equality through communication. Communication, in the context of cyberculture, evokes Norbert Wiener's definition of cybernetics as "control and communication in the animal and the machine."[62] In the introduction to *The Human Use of Human Beings,* he wrote that he defined *cybernetics* in those terms because the two are necessarily interlinked: "When I control the actions of another person, I communicate a message to him. . . . Furthermore, if my control is to be effective I must take cognizance of any messages from him which may indicate that the order is understood and has been obeyed."[63] He further argued that "society can only be understood through a study of the messages and the communication facilities which belong to it," and that communication between humans and machines would play an increasingly important role in the future.[64] This language evokes the historically racialized unfreedom that characterizes U.S. capitalist culture, and Wiener was especially concerned with ethical questions related to automation and cybernation, which I will return to later in this book.[65] The sense of communication as a form of control also resonates with the operative, materialistic, and consumerist sense, that it is given in the McCone Report. In the McCone Report, as critiqued through Marcuse, all human desires and forms of communication are subsumed into production for technocapitalism and the domination of nature. Black Watts is caught in this logic; it is undeveloped by so-called technological progress and thus not able to fully access the benefits of that progress, especially the signs of its commodity culture. However, at the same time, it is still forced to submit to its logic. Specific to Keynesian technocapitalism, this contradiction is part of a longer history of racial capitalist inequality.

Guy Debord, in "The Decline and Fall of the Spectacle-Commodity Economy" describes this double bind as the "universal" of the spectacle economy in relation to the Watts Rebellion:

> The American blacks are a product of modern industry, just like electronics or advertising or the cyclotron. And they embody its contradictions. They are the people that the spectacle paradise must simultaneously integrate and reject, with the result that the antagonism between the spectacle and human activity is totally revealed through them. The spectacle is universal, it pervades the globe just as the commodity does. But since the world of the commodity is based on class conflict, the commodity itself is hierarchical. The necessity for the commodity (and hence for the spectacle, whose role is to inform the commodity world) to be both universal and hierarchical leads to a universal hierarchization.[66]

Whether the Watts Rebellion was at its heart a rebellion against Debord's specific sense of exchange value and the commodity is debatable, given the extent and heterogeneity of racial antagonism in the United States. However, the passage captures the contradictions of race and capitalism in Watts. The system both demands the incorporation of impoverished Black Americans as consumer subjects that fully accept the "universal" value of commodity culture while simultaneously excluding them from full access to it. It is especially the supposed universality of commodification that is a part of the liberal capitalist culture that the McCone Report valorizes, that Purifoy, the Boggses, Marcuse, and Debord critique, and that in turn calls into question any mass politics focused on maintaining effective demand in the face of automated unemployment. This version of materialistic universality is a linear chronological one, where the consumer capitalist future is qualitatively like the consumer capitalist past, while the future negates the present in the ways it continually demands an investment in novel commodities. In this world, safety is premised, as Marcuse argues, on the military-industrial complex, which tries to contain not only Soviet and anticolonial challenges to the system but also domestic and Black ones.

To reiterate the obvious here, the consumer, or administered subject, is racialized as "White" and fully human, whereas to not have full access to commodities and safety is to be racialized as "black" and not fully human. For the McCone Report, to access safety and commodities, Black subjects must take on "White" habits and forms of communication through attitudinal training and education. This is impossible in McCone's world, but it establishes the normativity of what can be called "White" labor and consumption as the acceptance of the operational logic of this system.

Purifoy argues that for all humans, arts "education through creativity," which is to say not the "transient" forms of satisfaction found in the art object/commodity, is the "only way left for a person to find himself in this materialistic world."[67] While Purifoy rhetorically says that for all humans the materialistic world offers only "transient" forms of satisfaction, and he was clearly committed to the universality of the human, the values of the materialistic world are not available to Black residents of Watts. Today, *creativity* is a largely degraded term used to describe urban creative professionals or the marketing of "creative office space" in gentrifying city centers—uses of the term that emerge from neoliberal capitalism's counterrevolutionary colonization of the counterculture's ethos.[68] In the 1960s, *creativity* had yet to take on these connotations, and Purifoy's commitment to it as an antagonistic concept to the materialistic is connected to a longer tradition of African diasporic resistance to racial capitalism, especially in the Americas. This form of creativity insists on the (counter)universality of Black life and cultural practices. It especially resists the reduction of Black life to labor and value production for universal White capitalism.

In the self-published 1967 newsletter *One to One: Quarterly Report on Aspects of Creativity,* Purifoy explicitly distances his conception of creativity from productivity:

> In essence Art and Creativity
> are eons apart.
> Creativity
> is the act of doing
> void of the idea of productivity.
> Art is what people
> say about Creativity.[69]

In the McCone Report, education and communication are functional and goal oriented, reproducing a normative value of labor and communication as productive for the liberal capitalist state. Here, Purifoy explicitly calls into question the value of both productivity and the art object, and by implication its value as a commodity. The poetic fragment emphasizes the gaps between terms that might otherwise be thought of as synonymous: "art" and "creativity," "doing" and "productivity." These are categories, "what people say," that create false equivalences and eliminate difference. This happens especially temporally. Reducing creativity to art not only makes art operational but also eliminates the time, the "eons," when "doing" happens and simultaneously devalues the creative social person and reduces the present to the future art object. Purifoy's creativity opens up time and creates the future as a horizon not reduced to commodified forms of fulfillment.

As I argue, Black Power cybercultural theorists understood the conjuncture to be one in which forms of labor and social reproduction were being reorganized, which, while new, were connected to historically racialized distributions of waged and unwaged labor. Purifoy's junkers of the community were not outside of this but instead were involved in a contradictory informal economy, one harkening back to long-standing distributions of unequal labor while being at the leading edge of flexible forms dialectically related to high-tech employment. Thus, although not factory workers (or aerospace technicians), their labor does not stand apart from these other forms, and Purifoy's critique of productivity and conceptualization of nonproductive, self-valorizing work shares much with *Facing Reality*'s theories of working-class self-movement and creativity. Written in 1956 by C. L. R. James, Grace Lee Boggs, and Cornelius Castoriadis (who used the pseudonym Pierre Chaulieu) of the group Socialisme ou Barbarie, it was motivated by the Hungarian Revolution, but it was more broadly concerned with the rise of working-class autonomy in a period of automation. For them, this was a moment of crisis—one driven not by technology, the party, or management but by the self-movement of the working class that put them at the forefront of industrial, technological knowledge and of political change. Most important, they argued that the horizon toward which this self-movement through "internal antagonism" was directed was "**not** the enjoyment, ownership, or use of goods, but self-realization, creativity based upon the incorporation into the individual personality of the whole previous development of humanity. Freedom is creative universality, **not** utility."[70] For *Facing Reality,* creativity exceeds the valorization requirements of capital and points to all of those activities denied by the time of commodity production and consumption, as well as pointing beyond this to a reevaluation by racialized and working-class people of what constituted the human in a more-than-capitalist world.

Facing Reality's understanding of creativity was informed not just by the Euro-American industrial proletariat but also, as Christopher Taylor notes, by the creative resistance of enslaved and formerly enslaved peoples in the Caribbean and the Americas to the demands of the plantation.[71] Taylor examines the conceptual lacuna of antiwork Marxism, especially by Antonio Negri and Michael Hardt, which, while making important contributions to how neoliberalism seeks to make all activities value-productive for capitalism, fails to emphasize the foundations of antiwork praxis in Black and Caribbean emancipation movements. For Taylor, C. L. R. James is especially important because he not only theorized a "dialectical" understanding of the capitalist valorization process where "class composition produces an irreducible excess" that James called "self-activity," but also conceptualized capitalist temporality as "asynchronous, palimpsestical."[72] James, the Boggses, and their associates at *News and Letters* and *Correspondence* theorized not just the extent to which capitalist valorization exceeded the factory and entered all aspects of life but also the modalities of resistance to the "normativity" of work as a "goal-oriented social activity" where needs are satisfied only through commodities.[73] If antiwork Marxists often date these theories to the Italian autonomist and *operaismo* movements, those connected to James argue that resistance to the normativity of productive labor is rooted in the resistance of slaves and former slaves to the plantation economy. Or, as Sylvia Wynter wrote in the unpublished manuscript *Black Metamorphosis: New Natives in a New World,* written in the mid-1970s about the creation of a Black "New World" culture out of the Middle Passage and the plantation regime, "Labour time is not man's lifetime. Man's vital demand is the demand to participate in the constitution of a sense of self-worth by and through his creative action upon the world."[74]

Purifoy's antiproductive, creative reuse of junk is best located in the palimpsestic history of the cybercultural era and African diasporic struggles for freedom. This rehistoricizes and reframes his question about junk and the community of junkers in Watts—"what is the true value of these materials over and above sale to junkyards for a few cents?"—as it does the Boggses' own assertion that "now that man is being eliminated from the productive process, a new standard of value must be found. This can only be man's value as a human being."[75] In *66 Signs of Neon,* Purifoy strives to posit "man's value as a human being" by developing an aesthetic that asserts the universality of Watts through the creative reuse of impoverished materials available to the community—the junk that was, as Kellie Jones writes, "democratic; it didn't discriminate."[76] Purifoy saw the process of creativity and education not as simply "recreation" but as "rank[ing] alongside food and shelter as absolute necessities."[77] It is thus something fundamental—something related to the material histories of technocapitalism, the historic racialized distributions of waged and unwaged labor, the reevaluation of Black life in Watts, and the longer creative struggle for freedom against the normative demands of productive labor based on the plantation regime.

After *66 Signs of Neon,* Noah Purifoy became disillusioned with the potential of art to transform the world. He did not, however, become disillusioned with the creative process. For a time he returned to social work, and in 1976 he was appointed to the new California Arts Council, where he hoped to effect a transformation of values at the state level. Purifoy largely authored the council's program, writing, "The society itself is bent on a course of growth and consumption which seems to render all other human values—such as beauty—expendable. For sanity and survival, we feel the priorities should be reversed."[78] Ultimately, because of a combination of budget cuts and professionalization, Purifoy left the council in the early 1980s and returned to junk sculpture in the Mojave Desert. Like *66 Signs of Neon,* works like *Bessemer Steel,* which I opened this chapter with, question the values of growth and consumption, as well as the forms of expendability of technocapitalism and the racial Capitalocene; by doing so through the junk materials produced by capitalism and readily available in Watts, they reassemble histories and open alternative possibilities to these logics. The geologic time of melted lead and rusting steel must be creatively linked to the time of informal labor, as well as to the histories of a racial Capitalocene that, dating to plantation slavery, has distributed junk and people in common and in ways that are often only belatedly recognized. Following Purifoy, refashioning these materials is not a labor that is easily made reproductive for commodity culture. Instead, it calls into question processes of valorization, and it is this creative questioning and refashioning that points to both the longer trajectories of African diasporic creative resistance to a regime of value that reduces life to junk and to the creative making of new values out of the materials of racial capitalism.

4

Material Imperialism and the IBM Machine

PAULE MARSHALL'S *THE CHOSEN PLACE, THE TIMELESS PEOPLE*

Near the end of Paule Marshall's 1969 novel *The Chosen Place, the Timeless People,* Merle Kinbona, who is the Black heir of a former plantation in the fictional Caribbean country of Bourne Island and the intellectual and spiritual center of the novel, discusses family and futurity with Allen Fuso, a White anthropologist who is working on the island. In response to an inquiry by Merle about his statistical work, Allen says, "I can do this with my eyes closed. I'm a walking IBM machine, don't you know that? Always was, even as a kid. Mine's just one of those freak talents. But this isn't what I'm talking about. What I'm trying to say is that I've never done anything that was a real challenge, that didn't come easy like the stuff here."[1] Allen's self-definition as an "IBM machine" is representative of the ways that Marshall explores race, gender, and class through technology in the novel. This passage is also heavily weighted with Allen's suppressed sexuality and his unrequited love for Vere, another resident of the impoverished region Bournehills, who has just died tragically in a car crash. Allen is struggling with what he describes as his inability to be fully present in life, to which Merle suggests that he might return to the United States to marry (something Merle knows is not possible). He responds by saying, "A nice girl and some children, eh? . . . And why didn't you add while you were at it a nice job teaching at some nice Midwestern university like Michigan State or Wisconsin—that goes along with it, after all. Or better yet, a nice job making a lot more money in private industry doing research in cybernetics and the like. They're eager for guys like me, the computer minds. . . . A nice girl!"[2] In *Cybernetic*

Revolutionaries: Technology and Politics in Allende's Chile, Eden Medina notes that although cybernetics is nearly always described as a national phenomenon, in fact it has an underexamined transnational history.[3] This is the most significant of the small number of passages in which cybernetics or automation appear in *The Chosen Place, the Timeless People.* However, this novel is central to understanding Black Power cyberculture because of the way it narrates and theorizes how cybernetics as a political, imperial form was not simply a product of MIT or Silicon Valley but instead developed dialectically between the United States and places like the Caribbean.

The Chosen Place, the Timeless People focuses on the continued legacy of slavery and colonialism in the immediate postcolonial era in the context of a proposed but ill-defined United States–led development project that comes to the island with the intention of remaking Bournehills. The Center for Applied Social Research (CASR) is led by anthropologists Allen Fuso and Saul Amron, and the latter's wife, Harriet. Development is supposed to mark a break with the colonial past, but CASR is itself entangled with that past, and it is Harriet's family wealth, initially derived from the slave trade, that funds the project. As in development theory, technology is central to the novel, especially how the technologies of development were, and are, connected to histories of slavery, processes of racialization, and transformations in the planetary economy during the 1960s. Black Power generally, and Black Power cyberculture theorists explicitly, argued that Black revolution in the United States must necessarily be international.[4] Here, Paule Marshall reconfigures the geographies of Black Power cyberculture, and the encounter between White modernizers and the "timeless people" of Bournehills—Black, largely poor agricultural workers—is an opportunity to theorize through the techniques of realist narrative how race and class composition are connected to the histories of the plantation and then-new technologies of planetary capitalist value extraction. Marshall provides us a way of connecting the United States to the Caribbean and the African diaspora while raising ultimately unresolved questions about racialization and technology as well as what sorts of new forms of social reproduction politics are necessary to provide alternatives to neocolonial cybercultural inequality.

So far, each chapter of this book has explored how automation is connected to the plantation machine and the logics of technocapitalism understood as both a planetary and imperial phenomenon. For Grace Lee and James Boggs, these connections were marked by the difference between undevelopment in places like the Caribbean, which they defined in "The Myth and Irrationality of Black Capitalism" as "one-crop countries" organized to "supply raw materials or agricultural produce to Western imperialists" and who were struggling to "develop themselves industrially," versus Black Americans, who were undeveloped through industrialization.[5] Despite these differences, the Boggses, King, and in a different way Purifoy ask us to think creatively about how the cybercultural era in the United States was part of a planetary organization of technological value extraction and political responses to it. In this chapter, I turn to how technologies, whether cybernetic or otherwise, are explored in the novel through their role in the production of racialized, gendered, classed, and imperial subjectivities—whether this be Allen's claim that he is a White cybernetic machine incapable of heteronormative reproduction, Vere's relationship to American technologies of mobility like the car or the plane, the sugarcane presses around which the economy of Bournehills revolves, or the reproductive technologies of the home that largely dominate Merle's life, and that she must break from at the novel's end in order to imagine some form of diasporic futurity.

The novel's portrayal of these conflicting social and material forms, especially as they are linked to the emergence of U.S. imperialism, is the subject of Kamau Brathwaite's 1970 review of it. For Brathwaite, the novel's central concern is with narrativizing the complex ways that the Caribbean as "home" was enmeshed in a world of "materialism imperialism." For Brathwaite, *The Chosen Place, the Timeless People* was an exceptional novel that changed his "way of looking at West Indian writing."[6] According to him, before Marshall, "West Indian novels have been so richly home-centered, that they have provided their own universe" and have not positioned themselves either in relation to "communal history" or the "larger context of Third World underdevelopment." He continues: "The question however remains as to whether the West Indies, or anywhere else for that matter, can be fully and properly seen unless within a wider framework of external impingements and internal change. The contemporary West Indies, after all, are not simply ex-colonial territories; they are underdeveloped islands moving into the orbit of North American cultural and material imperialism, retaining within themselves stubborn vestiges of their Euro-colonial past (mainly among the elite), and active memories of Africa and slavery (mainly among the folk)."[7] I take up Brathwaite's line of thought in this chapter to examine how *The Chosen Place, the Timeless People* helps us see the cybercultural revolution as a moment of "material imperialism" in which cybercultural technologies not only organize forms of racialized and gendered un- and underemployment, as was so often the focus of U.S.-centered discussions, but also organized more intimate, personal, and domestic relations. Marshall called *The Chosen Place, the Timeless People* a type of "historical novel" because its form was capacious enough to explore the impact of the "four-hundred year history of the hemisphere" on the "relationships" and "politics" of characters from multiple racial and social backgrounds.[8] I would add to her description that the novel is arguably the Black Power–era text that most thoroughly explores the extent to which the cybercultural era was characterized by both a planetary history of plantation technologies and racialized and gendered anxieties about technological change.

Hortense Spillers says of *The Chosen Place, the Timeless People,* "Whatever Caribbean scene the novel brings to mind, we perceive Bournehills, the fictitious country of the novel, as a continuum of geopolitical and cultural moments surrounded by Marshall's order of the timeless. . . . Both the United States and the Caribbean are read, consequently, by way of the economics of captivity and their implications for the contemporary world."[9] As I explore in this chapter, what Spillers calls "the timeless" is primarily organized by the technologies that are left over from slavery and colonialism. By *technologies,* I mean all of those material forms through which life is organized to continue the plantation economy, whether that means the forms of capitalist value extraction mobilized for the export economy or the forms of social reproduction that defined the subjectivities of the Black inhabitants of Bournehills and White modernizers like Allen Fuso.

These include, for example, Merle Kinbona's guesthouse, Cassia House, where the American CASR team stays during their time on the island. Kinbona is London educated, and at the novel's start, she has recently been fired from her position as a high school teacher for teaching the history of Cuffee Ned's nineteenth-century slave rebellion. Cassia House is a former plantation founded by one of her distant relatives and passed down to her by her White colonial administrator father. This possession is weighted with historical and racial entanglements: Merle is the daughter of one of her father's Black servants, and he long ignored their relationship. Cassia House is introduced: "Rambling, run-down, bleak, the house was one with its surroundings, as much a part of the stark landscape of sea and sky as the sea and the dunes and the boulders strewn in the surf."[10] Like much of Bournehills, such as the Cane Vale sugar-processing factory, which dominates the region's economic and social life, Cassia House can be read as a ruin—in Ann Stoler's sense of ruins as not only decaying structures but also as "imperial formations" that are active processes linked to "what people are 'left with': to what remains, to the aftershocks of empire, to the material and social afterlife of structures, sensibilities, and things."[11]

Stoler's concept of ruination provides insight into the ways that the ordering of past, present, and future are bound to the material in *The Chosen Place, the Timeless People,* as well as how "rambling, run-down, bleak" material forms like Cassia House are not confined to the past but continue to be active in the present, influencing both processes of racialization and characters' sensibilities. The major task of this chapter is to analyze how the novel narrates the ways that those technologies left over from slavery and colonialism are connected to, and help to define, ostensibly more contemporary technologies like those of infrastructure modernization, cybernation, and information that in the period were primarily presented as future oriented and opening up new forms of thinking and doing. Alternatively, the novel suggests that as imperial formations, these new technologies are also repetitive forms, especially when examined in terms of the hierarchies of race, gender, class, and sexuality. Most important, the novel narrates and theorizes the centrality of thinking cyberculture from the Caribbean as an overlapping site particular unto itself but intertwined with the United States' imperial project.

In *The Sound of Culture,* Louis Chude-Sokei argues that because of the creole history of the Caribbean, its theorists have been particularly effective at untangling the history of race and technology as mutually constitutive concepts. Throughout the novel, the island's hybridity is treated as both a sexualized cliché and an important feature that allows characters, both Black and White, greater freedom of thought and interaction than would be possible in the United States, where racial lines were more strictly defined. This proves especially important when it comes to thinking the subjective features of cybernetics, which is presented as an imperial technology. However, it is precisely because it is discussed through Merle and Allen's interracial friendship that its (often veiled) implications as a technological formation can be explored.

The exploration of these political-technological issues takes the form of the realist novel. Although this is the only novel that I will be analyzing in *On the Eve of the Cybercultural Revolution,* it makes explicit a claim from the introduction. The cybercultural era is a conjunctural concept that attempts to name a technological reorganization of racial capitalist production and social reproduction and was thus at the time, and as I treat it from the vantage of the third decade of the twenty-first century, a periodizing and thus also a narrative category. Fredric Jameson argues in *The Antinomies of Realism* that the realist novel is best defined by a contradictory temporality between the pastness of the telling of the tale and the presentness of description (as well as the third time of the reading of the text).[12] As I have already begun to argue, this contradiction is explicit in *The Chosen Place, the Timeless People* in that it uses the realist form to explore the tension between the nonpastness of the technologies of slavery and the presentness of the technologies of U.S. imperialism. The novel exemplifies Saidiya Hartman's sense of the "time of slavery" that "trouble[s] redemptive narratives" about the ends of slavery and "negates the common-sense intuition of time as continuity or progression," which I discussed in chapter 1.[13] The realist novel form is in Marshall's hands especially effective at condensing the claim that the cybercultural era, and automation technologies more generally, must necessarily be thought in terms of the histories of the plantation. These repetitions and the apparent nonpastness of the past are explored most explicitly in the technological repetitions of the sugarcane press, the domestic technologies of the plantation, and cybernetics as a form where feedback seems to repeat, but in new ways, unequal social relations. At the same time, while conscious of the nonpastness of the past, the novel also considers what future and alternative forms of social reproduction might look like. These, however, are class based, and the future for Saul Amron, who returns to the cybernetic home of Stanford, is quite different from the Timeless People, who remain in Bournehills, and Merle, who is able to leave the island to attempt to reconnect with family and a diasporic future in Africa. Ultimately, however, it is the novel form itself that defamiliarizes the geographies and actors of the cybercultural era; in this way, it is a technology of social reproduction that works to open up histories as well as potential futures.

This chapter proceeds in three parts. Through the Cane Vale sugar-processing factory and Cassia House, in the first section I examine how the novel presents the plantation machine as both a productive and reproductive entanglement with the slave and colonial past that is resistant, in multiple ways, to the new technologies of development. The second section will look at how the development project is part of a new biopolitical and information-based regime of racial capitalist accumulation and how information and the plantation machine are the contexts that Marshall uses to explicate the racial, gender, and reproductive politics of Allen's cybernetic subjectivity, with which I opened the chapter. In conclusion, I use Katherine McKittrick's sense of Black women's geographies in order to consider how *The Chosen Place, the Timeless People* simultaneously imagines futures entangled with planetary technocapitalism and ones alternative to it.

The Plantation Machine

The continuing time of slavery and colonialism suffuses the postcolonial world of *The Chosen Place, the Timeless People.* The region of Bournehills is primarily the home of the Timeless People of the novel's title. They are often treated by the island's elite as mysterious and opaque, resistant to the commodity forms of postcolonial modernity. This resistance, however, is in fact an attempt by people who are the descendants of slaves, and still primarily poor agricultural workers who continue to labor in the afterlife of slavery, to win and maintain a degree of autonomy in the face of new attempts at dispossession and proletarianization. Timelessness thus speaks to the afterlife of slavery and resistance to it, as well as to the extent to which it is in fact a political-technological formation and a function of the commodity form more generally. Explicating how

Marshall writes the historical present of the technologies of slavery and colonialism is thus central to how she tells us these "past" technologies, as racialized and gendered forms, are constitutive of the "new" technologies of cybernetics and information.

The character of Vere is representative of the novel's technological problematic. It begins with Vere's return to Bourne Island by plane after a number of years in the United States as a guest agricultural worker. Vere, we are told, has spent the flight "stud[ying] a ragged well-thumbed manual on the repair and reconditioning of used cars"[14] and largely ignores both the view of the Caribbean outside the window and the CASR team members, who are the only other passengers. It is only as the plane begins to descend that he turns his attention to what is outside the window and readers get a sense of the geography of Bournehills and the ridge, Cleaver's High Wall, that separates the region from the rest of the island, which is described as resembling "a ruined amphitheatre whose other half had crumbled away and fallen into the sea."[15] Bournehills is thus geographically separated from the rest of the island, but it is also entangled with planetary capitalism via the technologies of modernity like the airplane, and the extraction export economy more generally.

The plane also offers a synoptic view of the verdant island that parallels the trajectories of development. Development is characterized by a linear temporality in which technocommodity culture, normatively modeled on countries like the United States, Japan, and those in Western Europe, requires the destruction of the environment and existing social relations in order to produce the new relations of development. As Gilbert Rist argues, this linear temporality is both a narrative form and a belief system that is continually insisted on and reproduced even when it fails to result in so-called development.[16] *Development* as a way of conceiving of international relations was first introduced by Harry Truman in his 1949 presidential inaugural address, at the height of Keynesian state capitalism, as a parallel to the Marshall Plan in Europe and as a U.S. intervention into decolonization that in many respects drew on the Monroe Doctrine. Alongside a claim about the place of the United States both as a capitalist imperial power and as a normative social form, development was to Truman, and continues to be to its proponents, a technological ideology.[17] As Michael Adas argues, "modernization theory" in the period was a way of "resuscitating the technological triumphalism" of World War II and a way to "elevate generalized claims regarding America's superior technical prowess and material culture" against the Soviet Union in the "contested zone" of the "Third World."[18]

Perhaps the most famous of these narratives of technological development is that of W. W. Rostow's *The Stages of Economic Growth.* Rostow's developmental arc moves from "traditional society" through "take-off" and "maturity," to the "age of high mass consumption," with the latter defined by Fordism, high GDP growth, and the fully organized commodity culture of the United States.[19] As the metaphor of "take-off" suggests, Rostow's stages are likened to an airplane (or spaceship) moving ever upward. Rostow's technological trajectories are paralleled and critiqued through Vere. The reader is immersed in Vere's technological world in which the plane's take-off from the United States carries both the member of the Timeless People most desirous of technological modernity and the CASR development team. However, what Marshall shows the reader in this opening scene of take-off is not a shiny new technological modernity but the ways the technologies of the past continue to structure the life of Bournehills. The region is described as being lush and green; however, the presence of the Cane Vale sugar-processing factory dominates the physical landscape and Vere's sense of place, just as it dominates the political economic landscape. For Vere, the sight of the factory from above instantly reminds him of the "deep pit" and the "rollers used to extract the juice from the canes" that also crushed and killed his great-uncle. The factory also has other personal connections; Cane Vale perversely represents home and impresses on him how "fixed and inevitable had been his return, how inescapable."[20] This moment represents a more general narrative tension in the novel. Vere is a figure of modernist desire who intends to use the technological knowledge he gained in spare moments in the United States in order to advance beyond the afterlives of slavery, as represented in the Cane Vale factory. Yet despite this, and despite the plane's function as both that which facilitates the movement of labor and capital from one place to another and a metaphor for development's taking off, the prominence of Cane Vale signals the nonlinear temporality that is in tension with linear technoprogress. Cane Vale and Bournehills' sugarcane fields represent a British colonial-era political and economic regime, which, although posited throughout the novel as outdated, continues to make a claim on the lives of Timeless People like Vere.

Cane Vale is always described in terms of the turning of its presses, which, along with the labor of the Timeless People, produces value but is also the source of injury and death. The single-crop sugar plantation, made by enslaved labor, is perhaps the most representative institution of capitalist modernity's emergence. As I wrote in chapter 1, Martin Luther King Jr. used the concept of the plantation machine to describe the ways Black Americans had been reified and dehumanized by technocapitalism. In the literature of the plantation period, the plantation machine was a functional metaphor because it was one part of a planetary capitalist machine that organized life, be it human or nonhuman, to extract value and produce commodities. It is for this reason that some scholars refer to the current era not as the Anthropocene but instead the Plantationocene.[21] In one of the few essays to examine technology in the novel, Justin Haynes argues that Cane Vale and its "rote and repetitive actions" are "symbolic of European Enlightenment in the Caribbean."[22] Vere is particularly important in Haynes's essay because his eventual death in an automobile he had rebuilt (and that ultimately turns on him and seems to purposefully fall apart) evidences the novel's concern with the "movement from the European Enlightenment to posthumanism," and a world of autonomous machines.[23] However, in the passages connected to Cane Vale (or Cassia House), the narrative of modernity is trapped in a contradiction: the plantation is not superseded but remains in a repetitive present. Development, the postcolonial, and the then-normative definitions of cybercultural revolution are all terms of transformation, whether marking a break with the past or a transition to new political and social forms. Marshall's treatment of Cane Vale suggests an alternative to this. If the plantation was a machine of synchronization with the main purpose of aligning the commodities produced by enslaved people with the export economy, then Marshall's novel brings out the ways that the plantation, and Cane Vale specifically, remains a machine that synchronizes time, racial hierarchies, and modes of violence in the immediate postcolonial Caribbean. It is necessary to consider this in depth in order to understand her specific critiques of development and cybernetics, which I will turn to later in this chapter.

The interior workings of the Cane Vale factory are seen for the first time through applied anthropologist and CASR project director Saul Amron on one of his data-collecting perambulations through Bournehills:

> [The factory was] a low sprawling building in serious disrepair, with a torn, rusted galvanized roof and a smokestack rising above it like a great phallus—he was reminded of the deep hold of a ship. There was the noise, for one—the loud unrelieved drumming and pounding of the machines that powered the rollers which crushed the juice from the canes, and the shrill, almost human wail of the rollers themselves as they turned in their deep pit. . . . And the light in the place was dim and murky as in the hold of a ship. . . . Because of the dimness and the cane chaff which came flying up from the roller pit to whirl like a sandstorm through the air, the men working there appeared almost disembodied forms: ghosts they might have been from some long sea voyage taken centuries ago.[24]

Cane Vale is presented here as the "hold" of a slave ship where contemporary laborers are also simultaneously the "disembodied" "ghosts" from "centuries ago," still bound to the plantation economy. Slavery, as Stephanie Smallwood and others argue, was a regime in which African diasporic peoples were racialized, categorized, abstracted, and turned into fungible bodies. As Smallwood notes, "The economic enterprise of human trafficking marked a watershed in what would become an enduring project in the modern Western world: probing the limits up to which it is possible to discipline the body without extinguishing the life within."[25] This mirrors Marshall's description of Cane Vale. The factory extends the disciplining regime of slavery into the 1960s-era present; the ghostly figures are never quite extinguished, and they continue to produce surplus value for Kingsley and Sons, the English company that owns Cane Vale. The passage also plays on Marx's formulation of the capitalist production process as being a relationship between "constant capital," like the factory, machinery, and land, and especially "dead labor" and "variable capital," or "living labor." Cedric Robinson, and others in the Black Radical Tradition, have troubled that binary, arguing that the "organizers of the capitalist world system" used Black labor power as "constant capital" during plantation slavery and in those regimes of forced labor that paralleled and followed it.[26] In their ghostliness, the Timeless People are representatives of the presentness of past regimes of value extraction and the ways that racial capitalism relies on socially differentiated, racialized life to make capitalism work by producing surplus value while abstracting life as value.

At the same time, the role of Cane Vale in *The Chosen Place, the Timeless People* also suggests an alternative to C. L. R. James's description of how the plantation organized labor power and revolutionary horizons in *The Black Jacobins,* which is also true of the League of Revolutionary Black Workers' conceptualization of Black labor and revolution that I will take up next, in chapter 5. James, writing in the context of the 1930s-era Trotskyist and labor movement, argued that the revolutionaries of Haiti were "workers," not simply slaves, and an "organized mass movement," because the "sugar-factories" had already made them "closer to a modern proletariat than any group of workers in existence at the time."[27] The continuing presence of the technologies of slavery and colonialism remains on the island, disciplining the Timeless People into laborers for the export economy. However, in the novel, they play a different role than the revolutionary proletariat. Instead, they are critical witnesses to the nonteleological nature of political change, but especially to the ways that technological modernity is, through the commodity form, a regime that demands their continual expenditure.

As Lyle Hutson, a lawyer who is part of the island's political elite (as well as Merle's former lover), says, the people of Bournehills are "behind god's back" because of their resistance to the commodities of development and modernity.[28] They are in fact something of a joke among the Bourne Island elite, a group that Marshall always describes in terms of both class and race. For example, at a party described early in the novel, Hinkson, a young lawyer "with a pale amber-colored face and the crimped dark-blond hair typical of the Bourne Island colored-whites as they were called," says to Saul,[29] "And the television set that British firm gave them for the social center played one day and then mysteriously broke down" and "The jukebox from America didn't last a week."[30] For the island's elite, who reproduce their class status through the ways they facilitate a dependent connection to planetary capitalism, there is something fantastical about the majority of the Timeless People's disregard for the technologies of affluence and spectacle. Yet as Leesy, Vere's great-aunt, says, technologies like cars and TVs were like a "new god" that demanded sacrifices—from Vere, from her husband who died at Cane Vale, and from so many others.[31] Instead of being an organized proletariat, the Timeless People are the novel's witnesses to the ways that the technologies of the plantation economy are technologies of violence. The forced obsolescence of the technologies of Rostow's age of high mass consumption does not signify in the novel maturity and freedom. They are instead a false freedom in which the permeable boundary between dead and living labor is congealed. The economy of Bournehills, and Bourne Island more generally, is what the New World Group, Caribbean economists, and theorists of structural dependence in the 1960s and 1970s that included, among others, Lloyd Best, Kari Polanyi, and Norman Girvan, called a "plantation economy further modified." They meant this in the specific sense of an economy that relies on the export of a single crop or commodity (and increasingly tourism), and that also relies on a dependent relationship to the metropolitan center for "technological change and taste formation."[32] Thus, for the Timeless People, space-age technologies of affluence not only signify dead labor but are also only accessible to them in a dependent form. It is from the critical perspective of the histories of the plantation economy that they see that the obsolescence of individual commodities and technologies, and not the commodity form itself, is tied to their own unequal relationship to that form.

Marshall's novel, however, does not ask us to see Bournehills and the Timeless People as strictly peripheral or dependent but instead in terms of a dialectic in which they are also central to the technoracial logics of capitalist modernity. From different locations, the Timeless People—and, as I will develop in the next chapter, the League of Revolutionary Black Workers—share a centrality to a planetary capitalism, past and present, in which labor is organized in terms of a plantation-factory regime that is necessarily racialized and gendered. Here, if not a revolutionary proletariat, the Timeless People do perform another kind of labor besides eking out a living working either directly for Cane Vale or by processing there what they are able to grow on their own small plots. As Louis Chude-Sokei argues, Black and African peoples have historically been called on to integrate machines and humans on behalf of the culture at large and to contain the anxieties of White people about becoming machines themselves. This was especially true during plantation slavery, and it continued to be so in the postslavery period, as represented in political-aesthetic terms like *primitive modernism*. On the one hand, Black peoples were seen as primitive and close to the earth, undisciplined, and ill fit for the White cultural world. On the other hand, they were seen as the most modern people, having become machinic or laboring automatons by being the people who were not just most thoroughly subject to industrial modernity but also the people who made it possible through their labor. The Timeless People are primitivized, and thus racialized, by the island's elite and others in the orbit of British and American colonialism (except for Saul and Allen, whom Marshall treats sympathetically) because they are not fully modern in the post–World War II consumerist, developmental sense. However, they do bear the weight of technological modernity to the extent that they are subject to the machinery of the plantation system and make the plantation economy work. Significantly, however, while they must work with Cane Vale in order to earn a living, they are explicitly skeptical of the extent to which they are called on to bear the weight of both colonial and post–World War II modernity.

So far, this should seem consistent with the ways that King, the Boggses, and the LRBW treat the plantation machine as continuous, if not synonymous, with cybernation and automation in the cybercultural era, especially in terms of its organization of labor and the ways that race is mobilized in order to extract value. Alongside this, Marshall narrates in depth something only referenced by those theorists. The cybercultural revolution is a neocolonial formation in terms of how it exploits the techniques of earlier racialized regimes of accumulation, in terms of its existence as part of a planetary economy that through the commodity form is universalizing and simultaneously particularizing in its uses of social and geographic difference, and in terms of the extent to which cybernation and automation are sociotechnical problems closely connected to so-called development and modernization.

Before moving on to discuss how this helps us read Allen Fuso as an IBM machine, let me address how Marshall explicitly connects Cane Vale to the domestic spaces of slavery and colonialism, like Cassia House. The New World Group called the plantation a "total institution" because it mobilized the entire population around the production of single crop for export.[33] In the novel, the technologies of the plantation are both productive and reproductive, and the complexity of this dynamic is explored most thoroughly through Merle Kinbona. Although a member of the Timeless People, Merle's belated recognition by her White colonial father means that her economic and social status diverges from theirs. Merle is still subject to the technologies of colonialism and slavery; however, her sociopolitical condition is much more contradictory, and it ultimately allows her a type of escape from the limits of Bournehills.

For example, Merle drives a "badly used" Bentley, formerly owned by the island's last British administrator, and in the opening pages, she is explicitly defined in relation to the "ruined but still regal" old car: "no longer young . . . [and] declin[ing] towards middle age."[34] This is not simply a comparison. Throughout the novel, Merle is shown to be entangled with and affected by the objects and technologies left over from slavery and colonialism that populate her life. These are less obvious than Cane Vale, but they are particularly important because they are formative of the domestic and the personal in ways that Cane Vale is not. This is the case with the "rambling, run-down, bleak" Cassia House, which is also her personal residence and a former second plantation home of her White ancestors.[35] It is on the shores of the Atlantic, described by Marshall as the site of "Diaspora" and "enforced exile."[36] The interior of her bedroom, normally hidden from the CASR team and other guests, is an intensified version of Cassia House in general. When Saul Amron sees it for the first time after Merle experiences a mental breakdown near the end of the novel, soon after the breakdown of Cane Vale and Vere's death, he describes her room as a "museum" of artifacts from slavery that "expressed her."[37] These include a prominently placed and detailed print of a cross section of "a three-masted Bristol slaver" showing in detail how slaves would have been held in the cargo hold during the Middle Passage, and Merle still sleeps in the "antique" bed of the plantation's founder, Duncan Vaugan.[38] Throughout the novel, these objects are akin to what Jane Bennett calls actants in a world of thing power that compose the ecology of Merle's life.[39] In acting on Merle, Marshall also tells us that they are technologies of social reproduction. Merle has a daughter who is living in East Africa with her father, and I turn in the final section of this chapter to how these relationships encode her sense of the future. Here, domestic space is a technology, an ensemble of structures, materials, and techniques, that is based on the organization of the planter's domestic hierarchy of race and enslavement. Merle is caught in the racialized organizations of the body in the afterlife of slavery through objects like the bed or the diagram of the slave ship.

However, she is opposed to them as well. In addition to her guesthouse, Merle earns a living by being a controversial teacher who tries to teach the histories of Black struggles for freedom despite the government's expectation that she teach a conservative history of colonialism. Alongside this, she is also a prominent observer-participant in Bournehills' celebration of Cuffee Ned during Carnival. Merle is, in Alys Eve Weinbaum's terms, a Black feminist philosopher of history in the sense that it is through both her explicit statements and the ways that she is presented by Marshall that we understand the contradictions of the production and reproduction of Bournehills culture as it is trapped between the past (both subject to it and resistant to it) and the future. The extent to which the domestic spaces of Cassia House are reproductive of the histories of slavery and colonialism are also necessarily connected to the organization of value extraction at Cane Vale. *The Chosen Place, the Timeless People* thus not only draws attention through the realist form to the necessary contradictions between past and present but also traverses the false binary logic of production and reproduction, economy and society. If, as Jennifer Morgan argues, the concepts of commodity value, economic and scientific rationality, and race emerged as an ensemble of issues with slavery, and especially its reproduction, Marshall argues that this interrelationship continues to be constitutive, if in different and new ways, at the moment of U.S. capitalism's attempts to incorporate the plantation economy into its own postcolonial regime of accumulation. This is one side of the cybercultural logic of the novel. Through its narration of these connections and contradictions, it provides us with a new geography of material imperialism and suggests the extent to which cybernetics in the figure of Allen as IBM machine is also necessarily developed out of both postcolonial development and U.S.-based technocapitalism.

The IBM Machine

I began this chapter with Merle Kinbona and Allen Fuso's anxious and veiled discussion about his Whiteness and divergence from reproductive heteronormativity through the figure of the IBM machine and the corporate cyberneticist, which significantly takes place at Cassia House, one of the main sites through which the afterlife of slavery is reproduced in the novel. Allen, IBM, cybernetics, and development signal the new form of (cybercultural) modernity that is emerging in the novel out of the British colonial past. Fredric Jameson argues in *A Singular Modernity* that *modernity* is a narrative category, one that is always an ideologically charged form of rewriting and periodizing history.[40] In *The Chosen Place, the Timeless People,* this periodizing resonates with his complementary argument that the realist novel is defined by a necessary contradiction between past and present. Marshall's novel is a thorough critique of both linear narratives of modernity and the ideologies that constitute it. Although characters and character development are certainly central to its project, this critique is primarily structural.

CASR is led by the sympathetic figure of anthropologist Saul Amron, who becomes Merle's lover. Saul sometimes sees Bournehills as part of an undifferentiated Global South, but Marshall presents him as someone who genuinely wants the people of Bournehills to be involved in the project, even if he is consistently perplexed by what he sees as the mystery of the region. He ultimately comes to openly question the role of development and modernization and argues that CASR should only institute a small-scale, locally controlled project. Like most of the scholarship on the novel, Lizabeth Paravisini-Gebert argues that the novel centers on two primary themes: "that of the importance of history in developing the national and personal identity required to transcend the legacy of colonialism and slavery, and that of the need to foster Caribbean economic self-sufficiency as the only way to achieve and preserve true independence."[41] Marshall is explicitly critical of the logic of development; however, Saul and Allen (although not Harriet, who remains faithful to racial and colonial dependence until her death) do not undermine the need to foster self-sufficiency. If Bournehills is the chosen place for development by both U.S. philanthropists and the island's political elite, then it is also the chosen place for resistance to this project (even if the future remains uncertain), and it forces Saul, especially, to become a critic of development's universal claims.

However, even more than an active rethinking of development, the CASR project never really takes off, to allude to Rostow. When Cane Vale ultimately breaks down, Merle critiques the CASR team for their inability to do anything about it: "After all, you're from a place where the machine's next to God, where it even thinks for you, so I'm sure you know how to repair something as simple as a roller. Machines come natural to your kind. Well, then, show your stuff and fix this one . . . fix it. That's the least you can do. Or is that asking too much? Perhaps all you can do is walk about asking people their business. Collecting data. And writing reports."[42] Merle alludes here to how computers had already come to be seen in the 1960s as having taken over humanity's ability to not just think but act, constituting a form of proletarianization whereby the knowledge to repair machines has been ceded to the machine itself. Yet the loss of one type of knowledge is in dialectical relationship with the accumulation of another type: the collection of data to "prepare the way" for a project still to come, and whose future is made uncertain when Harriet has it put on hold after finding out about Saul and Merle's relationship.[43] In this respect, CASR is less a modernization project in the infrastructural sense; it is instead closer to a biopolitical one. In 1976, Michel Foucault argued that, alternative to sovereign power focused on the right over death and individual bodies, biopower seeks power over life: "Biopolitics deals with population, with the population as political problem, as a problem that is at once scientific and political, as a biological problem and power's problem."[44] Locating this shift in the eighteenth century when Europe underwent demographic changes and early industrialization, Foucault further described biopower as a "new technology of power" that included "a set of processes such as the ratio of births to deaths, the rate of reproduction, the fertility of a population, and so on," which became "biopolitics' first objects of knowledge and the targets it seeks to control."[45] In *The Chosen Place, the Timeless People,* instead of actually building infrastructure or creating commodity markets, CASR primarily focuses on producing a "demographic analysis of the population of Spiretown [the largest town in Bournehills] by age, sex, occupation and the like."[46] It should come as no surprise, following Foucault's thinking, that the shift in the material forms of imperialism, from British colonialism to U.S. postcolonialism, would require that CASR first produce a body of knowledge about the population of Bournehills in order to institute any future project.

Here, the goal of development is to more thoroughly incorporate Bournehills into a specifically U.S. accumulation regime. The production of Bournehills as population thus brings them further into the fold of what Jonathan Beller has called "Digital Culture 2" and "computational capital." Like Jennifer Morgan, according to Beller, the Middle Passage and the slave auction block "reveal the imposition of digital metrics (like 'price') on bodies . . . with flagrant disregard for their person," and they laid the foundation for the consolidation of Digital Cultural 1's "universal aspiration" to assign "quantity" to "qualities."[47] This is not only a world of widespread technological value extraction but also one in which the racializing hierarchies of modernity suffuse the social world. In the CASR data-collection project, the novel suggests how technologies of slavery are carried over into the post–World War II material imperialist world of biopolitical governance and data-based, computational value extraction. The CASR project might then be reformulated to parallel Marshall McLuhan's observation that "IBM began to navigate with a clearer vision" once it "discovered that it was not in the business of making office equipment or business machines, but that it was in the business of processing information."[48] CASR never attains this clarity of vision before it is shut down, even if the true message of development is that production and social reproduction must be reorganized in order to put into place a new accumulation regime. *The Chosen Place, the Timeless People* argues, however, that structurally emergent forms of technological value extraction do not fully break from the past but are capable of capturing it and reorienting it toward new ends.

Marshall described the novel as being about the potentials of Third World politics and Black Power; it is, however, also an interrogation of how White Power remains structurally productive, despite the best intentions of the novel's characters.[49] This is true in Saul's struggles with development (although as Jewish, he is presented as a more liminal figure) and most explicitly through Allen. In him, Whiteness is shown to be a dehumanizing form of power, especially when connected to gender and sexual normativity, which prevents authentic connections with others. Marshall also argues that Whiteness is a technocapitalist formation that is continually evolving and that draws on past forms while programming new ways to live.

At the beginning of the novel, Allen's White middle-class subjectivity is described as the product of the technologies of post–World War II American domesticity: "All the various strains that had gone into making him . . . might have been thrown into one of those high-speed American blenders, a giant Mixmaster perhaps, which reduces everything to the same bland amalgam beneath its whirring blades."[50] The result of this is his suppressed sexuality, his unrequited love for Vere, and his inability to be fully present and live, to borrow from Hannah Arendt, the *vita activa.* To return to Allen and Merle's discussion, which takes places within the plantation-reproductive space of Cassia House:

> "I'm a walking IBM machine. . . . Always was, even as a kid. Mine's just one of those freak talents. . . . What I'm trying to say is that I've never done anything that was a real challenge, that didn't come easy like the stuff here."[51]

> "A nice girl and some children, eh?" It was said with a contemptuous snort. "And why didn't you add while you we're at it a nice job teaching at some nice Midwestern university like Michigan State or Wisconsin. . . . Or better yet, a nice job making a lot more money in private industry doing research in cybernetics and the like. They're eager for guys like me, the computer minds. . . . A nice girl!" He turned away.[52]

It is through his encounter with Merle, Vere, and the Timeless People that he comes to the realization that he has been made from birth into a White calculating machine in the whirl of a post–World War II suburban blender, which is itself tied to processes of racialization, technological mastery, and value extraction in neocolonial capitalism. The passage is weighted with existential dread because his Whiteness and sexuality exclude him from modes of life, in very different ways, in both Bournehills and Madison, Wisconsin. This encounter also places Marshall's examination of Allen as IBM machine at the intersection of two traditions. First is what Louis Chude-Sokei in *The Sound of Culture* calls "Caribbean pre-posthumanism." Anticipating currents like Donna Haraway's cyborg theory on the boundary breaches between humans and machines as well as humans and other forms of life, creolization and the Caribbean's history of hybridity is mobilized by theorists like Edouard Glissant and Sylvia Wynter to interrogate how forms of colonial power circumscribe what constitutes the racialized human. From this perspective, Allen and Merle's discussion is not fortuitously located at Cassia House, at the "eastern boundary of the entire continent" and the place in the Americas closest to "the colossus of Africa."[53] Instead, this is *necessarily* so because it is the location where the histories of Europe, Africa, America, and technocapitalism are most explicitly entwined.

Second, however, while discussions of cybernetics and race are made possible in the narrative by specifically Caribbean encounters, the novel's examination is closer to then-contemporary U.S.-based, but imperial, discussions about race and machines as binary formulations that are nonetheless reliant on each other. Allen and Merle's discussion seems in fact to be in dialogue with Eldridge Cleaver's essay, "Convalescence," from *Soul on Ice.* There, he argues that *Brown v. Board of Education,* as well as popular music by Chubby Checker, Elvis, and the Beatles, had begun to undo the "social imaginary" in the United States that separates "mind" and "body," coded as White and Black. Written in the style of 1960s New Journalism, Cleaver's essay navigates the complex boundaries between stereotypes and their material force in the world while also taking seriously the politics of White popular culture's appropriations of blackness. The essay is above all a technological one in which Cleaver argues that up until that point, Black people in the United States had been largely reduced to the body in order to make technocapitalist culture work, and White people had been reduced to being "bodiless Omnipotent Administrators and Ultrafeminines" who are fearful of their own desires for the body as represented by Black people.[54] He describes a world in which racial difference is constitutive of dehumanizing techomodernity:

> In the increasingly mechanized, automated, cybernated environment of the modern world—a cold, bodiless world of wheels, smooth plastic surfaces, tubes, pushbuttons, transistors, computers, jet propulsion, rockets to the moon, atomic energy—man's need for affirmation of his biology has become that much more intense. He feels need for a clear definition of where his body ends and the machine begins, where man ends and the extensions of man begin. This great mass hunger, which transcends national or racial boundaries, recoils from the subtle subversions of the mechanical environment which modern technology is creating faster than man, with his present savage relationship to his fellow men, is able to receive and assimilate. This is the central contradiction of the twentieth century; and it is against this backdrop that America's attempt to unite its Mind with its Body, to save its soul, is taking place. It is in this connection that the blacks, personifying the Body and thereby in closer communion with their biological roots than other Americans, provide the saving link, the bridge between man's biology and man's machines.[55]

Allen is close to Cleaver's "Omnipotent Administrator," a machine mind seeking connection to his body through Vere and Bournehills. These historically produced racialized roles are deeply troubling, but Cleaver seems to be optimistic that they can be overcome, in spite of the many increasingly powerful forms of technological dehumanization and the seemingly unreasonable demand that Black peoples be the primary force of overcoming racial technocapitalism's binary logic.

Cleaver's interest in the politics of race and technology would lead him to question the role of machines in class composition and lumpenization, which I return to in the final chapter. Here the dialogue I have set up between Cleaver and Marshall foregrounds the role of race in the production of the body politics of cybernetic technologies. It is also consistent with a broader concern with these issues at the time. N. Katherine Hayles argues that in the post–World War II period, and in the shift from the human to the posthuman, the body came to be thought of as "the original prosthesis we all learn to manipulate" and that "the posthuman view configures [the] human being so that it can be seamlessly articulated with intelligent machines."[56] Central to this shift in views was how information, through the work of, for example, Claude Shannon, came to be seen as a pattern that could be disarticulated from its "material substrates." Hayles argues that this move consolidated the liberal conception that the "subject possessed a body but was not usually represented as being a body."[57] However, Alexander Weheliye, whom Marshall and Cleaver anticipate, has critiqued Hayles for a view of the posthuman that "reinscribes white masculinity as the (human) point of origin from which to progress to a posthuman state," further arguing that Hayles's posthuman is "little more than the white liberal subject in techno-informational disguise."[58] Weheliye argues that this reinscription happens because of an occlusion of the imbricated histories of slavery, the body, and racialization. At the same time, Weheliye contends that because of this history, "Afro-diasporic thinking has not evinced the same sort of distrust and/or outright rejection of 'man' in its universalist, post-Enlightenment guise as Western antihumanist or posthumanist philosophies. Instead, black humanist discourses emphasize the historicity and mutability of the 'human' itself."[59]

In *The Chosen Place, the Timeless People,* Marshall is more skeptical, at least in the passage under consideration, than Cleaver that racialized technopolitical binaries might be productively overcome. Whiteness here is also not so much the point of origin of progress but is rather marked by an impossibility, either for some authentic humanness or for some less authentic, posthuman liberation, explicitly because Allen is not an example of normative universal masculinity. The novel also adds an important feature to the debate staged here between Cleaver, Hayles, and Weheliye because despite the asterisks attached to Allen's gendered and sexual normativity, his cybernetic anxieties are connected to the forms of racialized neocolonial value extraction represented in Cane Vale, the plantation economy, and CASR's data-collection project. The novel shows that Allen's involvement in this project makes him part of a broader social machine and distribution of economic power, where the residents of Bournehills are transformed into data for potential value extraction for U.S. development, and thus forced to continue to live as dependent black labor in a plantation economy. Allen's resistance to becoming an IBM machine is thus connected to racialized and classed colonial dependence, although this does not free him existentially but only bodily from certain forms of manual labor.

Chude-Sokei argues that these issues have always been present in science fiction, if not in the realist novel, as well as in cybernetic theory. While the connections between race and technology in both literature and cybernetics have largely been unacknowledged (except in creolization theory), in Norbert Wiener's thought, the close connection between cybernetics as a theory of communication and control to actual existing histories of White Power, slavery, and capitalism in the United States was always present.[60] This is most explicit in his statement, "Let us remember that the automatic machine, whatever we think of any feelings it may have or may not have, is the precise economic equivalent of slave labor."[61] In terms of *The Chosen Place, the Timeless People,* perhaps most strikingly, in *The Human Use of Human Beings* Wiener equates the corporate and academic heads who seek to utilize technology and cybernetic ideas in order to increase efficiency and mass production to White supremacy, fascism, and antidemocratic organizations of everyday life. For Wiener, the ramifications of this are profound: "Those who would organize us according to permanent individual functions and permanent individual restrictions condemn the human race to move at much less than half-steam. They throw away nearly all our human possibilities and by limiting the modes in which we may adapt ourselves to future contingencies, they reduce our chances for a reasonably long existence on this earth."[62] Alternatively, Wiener argues that cybernetics is in fact a way of thinking that is potentially much more like life itself, in which "its present is unlike its past and its future unlike its present. In the living organism as in the universe itself, exact repetition is absolutely impossible."[63] Thus, although cybernetics as a political formation risks being captured by White corporate power in order to mass-produce a fascism modeled on the racialized history of slavery, an alternative, true cybernetic politics is open to new horizons of the future and organizations of life. This is certainly desirable, and it is this aspect of cybernetics that made it so appealing to the counterculture.[64] Marshall, however, gives us pause (and Wiener's worries about centralization and the bad uses of cybernetics are even more apparent in Silicon Valley today). Allen is an ethical political figure, and he struggles against the logics of corporate development. Despite this, cybernetics and data are nonetheless presented as repetitive technologies that do not move forward in time and involve qualitative change, but instead repeat the subjectifying and extractive logics of (neo)colonialism and slavery, although in new circumstances. Despite countercultural politics, and also in response to them, the cybercultural revolution as an imperial formation thus remains normatively reproductive of racial-, gender-, sexual-, and value-extractive relations, according to *The Chosen Place, the Timeless People.*

Divergent Futures

Ultimately, Merle and Allen's brief discussion does not fully explicate the relationship between IBM machines, cybernetics, and processes of racialization and the production of normativity in planetary capitalism during the cybercultural era. Despite this, Paule Marshall's narrative intervention demands that we fully consider the unexpected overlapping histories of plantation modernity and the emerging cybernetic distributions of life and value, and especially the ways that questions of the human were socially distributed in the Caribbean as part a U.S. material imperialism seeking to capture the plantation economy further modified at a moment of Third World struggles for independence and domestic Black Power struggles for freedom. Above, I wrote that Weheliye critiqued Hayles's omission of race in her discussion of posthumanism. Despite this, Marshall shares with Hayles a belief in the importance of literature to understanding technological histories. Hayles argues, "Literary texts often reveal, as scientific work cannot, the complex cultural, social, and representational issues tied up with conceptual shifts and technological innovations. . . . It is a way of understanding ourselves as embodied creatures living within and through embodied worlds and embodied words."[65] *The Chosen Place, the Timeless People* thus not only asks us to follow the specificities of race, technology, and gender's imperial intersection but also to use the novel to theorize these relationships in new contexts, as I have tried to do here in relation to Black Power cyberculture. The novel form also allows Marshall to speculate on the potential futures of its characters, which diverge especially on the basis of their class position. At the end, she gives us five trajectories of the future. The first, and least prominent, is that of the island's elite, who remain committed to the plantation/export economy and increasing tourism. Second, Allen decides to stay in Bournehills primarily because he plans to continue with CASR even if the future of the project is uncertain. Despite his intention to stay, his personal status remains unresolved because his Whiteness is still constituted through the continued possibility of becoming a middle-class machine based on a data-extractive regime of accumulation, even if here the IBM machine has a problematic relationship to heteronormative futurity. Third, Allen must necessarily be posed against the Timeless People, who remain Black and thus not fully human, according to the standards of development, because of their supposed incapacity for full technological incorporation into the new regime of accumulation. Yet they also remain engaged in the struggle to make new lifeworlds in the afterlife of slavery and colonialism and remain witnesses to the trajectories of Black freedom.

Fourth is a form of circular time. The last we see of the Timeless People, characters like Leesy, Delbert, and Stinger are saying goodbye first to Merle and then Saul. The novel ends where it began, with Saul on the way to the airport and the first of the seasonal storms on the verge of washing out the Westminster Low Road, which connects Bournehills to the rest of the island, and whose state of disrepair had initially prevented Merle from meeting Saul, Allen, and Harriet. In this respect, the temporality of the novel is one of repetition and circularity, like that of Cane Vale and corporate cybernetics. Saul, more than Allen, also seems committed to development—as well as to repeating his personal trajectory. Although Saul tells Merle that he is "more than ever convinced now that that's the best way: to have people from the country itself carry out their own development programs whenever possible. Outsiders just complicate the picture," he intends to return to Stanford—an unlikely place to put these ideas into practice because it was one of the centers of cybernetics and of the emergence of corporate information and data.[66]

Ultimately, what political lessons are we to learn from the seeming timelessness of the chosen place as it intersects with planetary technocapitalism? How can it help us to think about Black Power cybercultural politics? In conclusion, I believe the novel form itself, as an example of what Katherine McKittrick calls Black women's geographies, to be the most helpful way of approaching the novel's politics. Bournehills (and Bourne Island in general) is what McKittrick, in conversation with Sylvia Wynter's work, calls a formerly "uninhabitable" zone.[67] These originally Indigenous spaces, like the Caribbean, were central to the ways that the genre of the human Man was created, and they have been made habitable for *homo oeconomicus* by turning them into workable lands for the commodity form, economic growth, and regimes of managing racial and gendered difference against which the normal is measured.[68] This is done through the technologies of slavery and (neo)colonialism, like the sugarcane press, the plantation, the extractive-export economy, and the institution of development as the linear emplotment of the technologies of consumer capitalism. As McKittrick argues, racism and sexism are central to the social production of space and that racial dispossession, and the genre of the human called Man2 is based on capitalist regimes of ownership, of which chattel slavery is one of the founding forms. Black geographies and especially Black women's geographies, she further argues, rupture normative forms of knowability and show that geography is not just space that contains social relations but is produced with them. Significantly, in terms of the novel, Black geographies are also struggles that think differently about ownership.[69]

The Chosen Place, the Timeless People narrates the historical present of Bournehills as a site of struggle to further consolidate the Man2 of international development, as well as the struggle against it to create a new world. The novel tells us that this struggle does not always take place as an explicitly linear and unambiguous trajectory of revolutionary overcoming but instead in in-between, complex ways. Bournehills is one site in the cybercultural revolution whereby development seeks to capture the older technologies of slavery and colonialism through the new data-oriented technologies of racial and population management and ultimately new infrastructures and expansions of the commodity form. The Timeless People are highly skeptical of the benefits of development and are instrumental in pushing Saul and Allen's ambiguous rejection of this new ownership regime and taking the first steps toward their (impossible) distancing from Whiteness as a normative property regime. As I've already noted, however, Allen's own struggles for freedom are necessarily contingent on the possibility of new forms of cybernetic value extraction.

Finally, Merle's trajectory of freedom is also ambiguous. It is both an escape from the material imperialism of home and an anticipated return to it that is in dialogue with the Timeless People, as well as with Allen. Critical of the technologies of slavery and development and cognizant that they have shaped her in dependent ways but have also afforded her a degree of wealth and freedom that separates her from the Timeless People, Merle ultimately decides to break from them to temporarily leave home in search of her estranged daughter and husband in East Africa. As she says to Saul at the novel's end: "My traveling papers. . . . I had to sell almost everything I owned to raise the money, but I managed. You should see my room. It's as bare as a bone. Everything gone—all that old furniture and junk. . . . And the wreck of a car is gone. . . . Gone!"[70] In these lines, Merle acknowledges and rejects what she calls the technological "junk" of colonialism and slavery. This is not a rejection of history but an attempt to reckon with the ways its materiality infuses every aspect of her life, and by implication the life of Bournehills, including the continued commodity value of the museum objects of Cassia House in the postcolonial marketplace.

This break can also be read as a break with the masculinist trajectories of development as well as the nation-state form in favor of diasporic freedom struggles because it is Merle and not members of the island's male political elite, like Lyle Hutson, who break with the regimes of economic dependence. *The Chosen Place, the Timeless People*'s economic politics are thus connected to the novel's formal politics. Belinda Edmondson argues in *Making Men: Gender, Literary Authority, and Women's Writing in Caribbean Narrative* that for male authors, the making of novels in the pre- and postindependence anglophone Caribbean was closely connected to the "making" of themselves as men through "the cultural authority that [had] been passed on to them from British intellectual discourse of the nineteenth century."[71] Thus, for Caribbean, and often diasporic, women like Marshall, it was necessary to "revis[e] what constitutes literary authority" as well as the masculinist and nationalist models of the novel.[72] In Merle's break with the material imperialism of Bourne Island, the novel makes a claim that it is through Black women's struggles that the geographies of freedom are best known. Further, if the dependent neocolonial nation-state is a form made habitable for Man2, then-new forms of Black freedom must be made in international or postnational spaces that are themselves made through the diasporic trajectories of women like Merle.

Yet *The Chosen Place, the Timeless People* does not claim that this is a clean break or politically unambiguous. As Merle says before leaving, "But I'll be coming back to Bournehills. This is home. . . . Sooner or later [one] has to take a stand in the place which, for better or worse, he calls home, do what he can to change things there."[73] Merle's trajectory of freedom simultaneously separates her from the rest of the Timeless People because her inheritance facilitates it, although she also remains bound to Bournehills as the chosen place of struggle over racial capitalist, technological modernity. At the same time, Merle's struggle inverts Allen's. She is headed to Uganda, where her daughter and ex-husband live. Ketu is an agricultural economist at Makarere University who left her after she had a relationship with a wealthy White woman in London. Merle thus seeks a diasporic heteronormative futurity that is at least in part connected to modernist agriculture as a technology that is essential for postcolonial independence. This would seem to oppose (while keeping in mind the racial, gendered, and national differences between them) Allen's rejection of the heteronormative family and the spaces that make it possible. The cybercultural politics of the novel are thus ambiguous in productive ways. On the one hand, it is necessary to break from the linear trajectories of development that use data, and potentially cybernetics, in ways that repeat the technologies of slavery and colonialism. On the other hand, according to *The Chosen Place, the Timeless People,* freedom struggles are messy, and their forms of futurity and social reproduction continue to be entangled with the past even when new horizons are opened—a problem that Marshall and Black Power cyberculture tell us must be continually reckoned with in the struggle for just, more-than-capitalist futures.

5

"DRUM Would Like to Welcome You to the Plantation"

THE LEAGUE OF REVOLUTIONARY BLACK WORKERS' EXPANDED THEORY OF PRODUCTION

The "Housekeeping Assignments Notebook" in the General Baker papers at the Detroit Revolutionary Movements Records is an approximately twenty-page notebook primarily containing, as the title suggests, assignments for chores in an unspecified household that include cleaning the kitchen, purchasing coat hangers, fixing the porch, and snaking the drains.[1] Dated February 29, 1969, and February 16, 1971, thus spanning almost exactly the existence of the League of Revolutionary Black Workers, it also includes outlines for meetings on various topics: "method of work," "centralize responsibility," "collective committee study." The notebook gives little indication of what went on in these study sessions, but there are two hints. On the first page is written "(1) What are the principle contradictions of imperialism? How have their cont. been affected by the emergence of social imperialism in the Soviet Union—Why did Russia become the home of Leninism?" At the top of the last, otherwise blank, page is "Ludwig Feuerbach & The Outcome of German Classical German Philosophy by Frederick Engels," which was written in 1886 and also included for the first time Marx's "Theses on Feuerbach" as an appendix. There is nothing in the notebook elaborating on these; however, they point to some of the League's main concerns: imperialism, Leninist political strategy, and the history of Marxism. More generally, although the notebook does not elaborate on the social dynamics of housekeeping, and first names are only rarely attached to any particular activity, it does suggest that in some ways, members of the League sought to think politically about all of the activities, including in the home, in which they were involved, even if gender relations within the League have been justly criticized.[2]

I begin this chapter on the League of Revolutionary Black Workers with this surprising document because of all of the Black Power cybercultural theorists, the LRBW was the most resolutely focused on Black workers and the point of production instead of on the ways that automation was increasing unemployment and thus required thinking in new ways about work and activity. This focus was in part due to the League's location in Detroit. In a period when the automobile industry employed one out of every six people in the United States, many shared the opinion of the unnamed Chrysler executive in the League's 1970 film *Finally Got the News:* "As the automobile industry goes, so goes the city of Detroit, so goes this country."[3] This was also a period when Detroit autoworkers had become majority Black and thus were, as League member John Watson says in *Finally Got the News,* "essential and key to the continual operation and the continual smooth functioning of a highly industrialized, highly complicated machine." Consequently, "we can use that power if we can ever get ourselves organized well enough to destroy that machine or to take it over."

Despite this, as the journal *Leviathan* said in an interview entitled "Our Thing Is DRUM!" with members Ken Cockrel and Mike Hamlin, "People who aren't familiar with the Detroit scene have an image of the League as only organizing at the point of production. Yet when you come to Detroit you see that the League is much more broadly based: you're into a whole bunch of other things," especially community organizing.[4] In this chapter, I ask how the League's theorization of Black labor at the point of production was interrelated with its cultural productions and conception of community. As I have argued throughout this book, the cybercultural era was seen as a moment, to borrow Nancy Fraser's phrase, of *boundary struggle* between production and social reproduction in which automation and cybernation were undermining Fordist–Keynesian's ability to reproduce itself. Despite their primary focus on assembly-line production, the "Housekeeping Assignments Notebook" intriguingly suggests that the League's theoretical perspective was much more expansive and verged on a theory of how social reproduction included not only factory relations but also how the family, home, and the education system were imperial, racialized, and capitalist formations. Yet the League never fully got there. Instead, I argue in this chapter that the League rethought the assumptions of Marxism at a moment of technological change in which they developed an expanded sense of *production* that included not only the assembly line but also an examination of the (re)production of racial capitalism in the community, as well as in educational institutions. Central to this theory of production was the League's extensive cultural productions, like the film *Finally Got the News,* the citywide newspapers *Inner-City Voice* and *The South End,* factory-based papers like *drum, elrum,* and *F.R.U.M.,* and student papers like the *Black Student Voice.*[5] As I examine below, through their editing practices, the League, as a collective author, made connections in these often intertextual publications between the factory and the community and examined how race was made productive for mediating the contradictions of cybercultural capitalism.

Too often, the history of the League has been framed in terms of its failures, whether this be member Ernie Allen's argument that the League broke apart primarily as a result of problems in its organizational structure, or Dan Georgakas's argument that *Finally Got the News* was unable to properly connect the relationship of the primarily Black men it focuses on at the beginning of the film to White workers, Black women, and the community that are the focus of its second half.[6] Perhaps the most famous statement on the League's failure is Fredric Jameson's discussion of cognitive mapping in *Postmodernism, or, The Cultural Logic of Late Capitalism.* Jameson contends that although *Finally Got the News* and Georgakas and Marvin Surkin's definitive history of the League, *Detroit: I Do Mind Dying,* "triumphantly survive" as "representation," they do so as "narrative[s] of defeat" that point to the spatial challenges of organizing a political movement against late capitalism.[7] That the failure of the League to take state power ends *Postmodernism* points to the importance of the League to the politics of the 1960s. In this respect, Jameson's book suggests—like Peter Linebaugh and Bruno Ramirez's essay "Crisis in the Auto Sector," about wildcat strikes in the automobile industry—that the ultimate success of the League, alongside its example for future movements, is that capitalism met its challenge through restructuring and a move toward "flexible" accumulation strategies.[8] Because of this, the end of the League is best understood in terms of counterrevolution and not failure.

However, more important, in this chapter I follow the examples of Errol Henderson, Jordan T. Camp, Fred Moten, and Morgan Adamson, who argue that the League was especially successful at both theorizing its political struggle against capitalism and also connecting that theorization to forms of organizing that took into account the specificities of post–Rebellion Detroit as well as the long history of Black working-class struggles.[9]

Regardless of one's evaluation of the League's political project, all this suggests the extent to which the LRBW was a multifaceted organization that involved factory and community organizing, a wide array of cultural productions, and capacious theoretical practices—or, as John Watson puts it, they were a "conglomeration of activities."[10] Founded in early 1969 before splitting apart in mid-1971, the League was a semiautonomous umbrella organization that sought immediate improvement to factory conditions; however, it also had long-term plans to organize a general strike led by Black workers to take local, and ultimately state, power.[11] It included revolutionary unions like the Dodge Revolutionary Union Movement (DRUM), which was founded in early 1968 to contest speedups and racist labor practices at Dodge's Hamtramck plant, as well as the Eldon Avenue Revolutionary Union Movement (ELRUM) and the Ford Revolutionary Union Movement (FRUM), to name only three of the associated groups. In addition to the factory-based papers that the RUMs published, the League produced its own citywide paper, *Inner-City Voice* (founded in 1967) and for a time ran the Wayne State University student paper, *The South End,* meaning that industrial organizations and an array of print media played equal, and overlapping, roles in the League. As Georgakas and Surkin write, the League "never functioned with the precision of a capitalist corporation or a Marxist–Leninist party. Rather than having a strict hierarchy, the League was organized into several components which had specific areas of work. Each component had a semi-autonomous character which often reflected the particular personalities of the individuals in charge."[12]

As they note, the divide between factory workers and nonfactory workers, like Watson and Cockrel, was often a source of tension. In this chapter, despite the prominence of those two, along with Luke Tripp, Mike Hamlin, and members of the Newsreel collective, I treat the League primarily as a collective author that included the often unattributed writers who produced the factory papers, and I argue that the League's thought emerges through the interplay of ideas across their intertextual media productions. In doing so, I examine the League's best-known works, like *Finally Got the News, drum,* and *The South End,* as well as interviews like John Watson's "To the Point of Production" and "Black Editor," as well as still largely unexamined texts like *Black Student Voice* and internal documents like "The History and Derivation of the League of Revolutionary Black Workers." Although formally different, in all of these, the League connects the demands of organizing in 1960s Detroit, the exploitation of Black labor on the assembly line, and analyses of how race is productive for cybercultural capitalism to the history of Black workers and their long-term political struggles.

Throughout, the League was especially involved in rethinking the assumptions of then-mainstream Marxism as it related both to organizing Black workers and the centrality of racial difference to capitalism. In "Black Editor: An Interview" that appeared in the journal *Radical America,* John Watson argued explicitly that a Marxist–Leninist and socialist approach was a necessary part of the Black freedom struggle. He counterposed this to Stokely Carmichael's then-recent claims that Marxism and socialism "were not for black people," as *Radical America* put it.[13] As he and the League would do elsewhere, Watson defined their position as a worker-based departure from "cultural nationalist," lumpen, "separate state," Black capitalist, and "haphazard" "revolutionary" positions that were dominant in the Black Power movement at the time. Yet though he argues that Marxism–Leninism was key to understanding capitalist production and anticapitalist organizational forms, he does not do so uncritically. Instead, Watson emphasizes that Marxism is a "particular method" and that the "Marxism of the past" did not have "all the answers." He argues that we must be attentive not only to the historical processes of racialization in planetary capitalism but also to the contingency of counterhegemonic struggles.[14]

In a January 1970 speech at a conference with Robert Williams and Emory Douglas organized by Newsreel, Ken Cockrel defended the League's record of keeping their organizers out of prison—unlike (implicitly) the Black Panthers. Embedded in this speech, Cockrel, a lawyer and future Detroit city council member, argued, like Watson, for a capacious sense of Marxist thought. As the League often did, he stated that "the point of greatest vulnerability" in a capitalist system based on imperialism and racialized labor was the point of production. However, Cockrel also says that when we talk about the "capitalist, imperialist machine," "we do not simply define workers in the orthodox sense of he who toils laboriously with his hands over a lather or on the line. . . . We say that all people who don't own and rule and benefit from decisions which are made by those who own and rule are workers."[15] He extended the definition of workers in particular to Black students. Ursula McTaggart argues that although the League primarily emphasized assembly-line workers in its politics, for them the term *worker* "became slippery," and because all of Detroit was associated with the auto industry, the League "identified auto workers with Detroit's entire black community."[16] Here Cockrel not only expanded the definition of worker but also the definition of the means of production and the "productive relations of this society."[17] Education, and other state apparatuses, he argued, determined not just whether one was or was not a worker, or will or will not become one, but the "means of production" are "organized" "in such a way as to make them unnecessary."[18] Workers must thus be organized not only to take control of the factory but in order to take state power to control the institutions through which the racialized and hierarchical social relations of capitalism can be transformed. Cockrel's expanded sense of production thus extends the concept beyond the factory and into the social relations of everyday life.

It is with the League's expanded sense of the production that I am primarily concerned here. It is by questioning what is meant by *production* that the League made its most important contribution to Black Power cybercultural thought. Contrary to even factory workers like James Boggs, who argued that the cybercultural revolution was increasingly producing outsiders who would be the new revolutionary force in society, the League argued that factory automation was not in fact taking place. Instead, automation in the auto industry relied on intensified Black labor—or, as Mike Hamlin put it, there was a reason "why those departments [referring to both automated and the most toxic sectors of the factory] are overwhelmingly black in every instance."[19] In focusing on automation's intensification of labor, instead of its capacity to produce unemployment, the League is part of a tradition of debate about technological unemployment that dates to the beginning of capitalism. This also makes their ideas particularly useful for understanding how automation requires intensified and most often racialized labor in contemporary workplaces as well—like, for example, Amazon warehouses.

In documents like "The History and Derivation of the League of Revolutionary Black Workers," the League examined both the history of racialized labor in the United States as well as the history of U.S. Marxist thought in order to theorize the contemporary factory, or "plantation," as a mode of production that required racialized labor to manage the contradictions of the assembly line. In doing so, they also challenged then-normative theories of class composition in the United States and argued that race prevented the proletarianization of White and Black workers. Ultimately, they sought to use this in order to better understand how a revolutionary politics could be organized in the cybercultural era. In the first part of this chapter, I examine their theory of Black labor on the so-called automated assembly line, as well as how they thought with and against a Marxist tradition of factory labor in technocapitalism. I turn to especially György Lukács and Karl Marx's thoughts on the factory as a model for society, the commodification and rationalization of labor, and the ways that automation required the total mobilization of science and nature in order to think about how the League's understanding of how race is productive for automation is a necessary component of this tradition.

In the chapter's second and third sections, I look at how the League's expanded sense of production involved the central role that cultural productions, or the "news," played in their collective work. In addition to being organizational tools, it was through these that the League produced itself as a collective and mediated its many constitutive differences. In my examination of *Finally Got the News* and *The South End,* I look at how editing was their primary form of production and how, through intertextuality, they further examined how race was productive for surplus-value extraction. Editing was also a technique that allowed them to make connections between the categories capitalism needs to keep separate, like race and class, and factory and community. Editing, which drew attention to how the League actually produced its media, was thus also a way of producing Black worker consciousness in order to produce revolutionary change. I end by turning to an analysis of *Black Student Voice* and the prose poem "Poetics" by Bobby Jean Cummings, which appeared in the second issue of *drum,* to examine how the League's expanded sense of production conceptualized the reproduction of the social relations of racial capitalism.

Automation in the Factory-Plantation

In volume 1, no. 3, of *drum,* the "New Employee" column begins by saying, "DRUM would like to welcome you to the plantation." Describing itself as the "Voice of Black Unity at Hamtramck Assembly," the "Racist Arm of the Chrysler Corporation," it tells new employees to "look around" to see that Black men and women have the worst jobs and are constantly under White management's surveillance. It ends by calling all new Black employees to the DRUM cause: "and now that you have been hired we feel that you are part of this struggle for black dignity."[20] Designating Hamtramck a "plantation" was far from a sensationalist rhetorical move, although it was also that, and *drum* (and the other RUM papers) were often criticized for their rhetoric, like calling Black employees that disagreed with them Uncle Toms, and for their sometimes denigrating descriptions of women.[21] Instead, defining the automobile factory as a plantation was constitutive of the League's theorization of capitalist production during the cybercultural era.

Alongside being an organizational paper, *drum,* in its totality, could be best described as a text that makes an argument for why Black workers (and others) should see it this way. These include especially the many pages describing the bodily harm and toxicity that the workers were subject to, which led the League to file a grievance with the National Labor Relations Board against both Chrysler and the UAW for failing to enforce safe working conditions.[22] Jonathan Flatley has described these journalistic practices as creating a "revolutionary counter-mood" in the tradition of Lenin that defamiliarized these conditions, allowing them to be seen in new ways by the workers who experienced or might experience them, thus facilitating action.[23] The very first page of *drum* describes the conditions at Hamtramck and why a wildcat strike on May 2, 1968, initially begun by White women, was necessary. It also claimed that photography was used "as evidence against some of the pickets and were instrumental in the discharge and disciplining of certain workers who took part in the walk-out and picketing," including, most prominently, General Baker.[24] The wildcat strike, which had a prominent place in worker self-organization and activity during the cybercultural era, was here an important collective action that Black workers used to defy the racialized disciplinary regime and the relations of intensified production on the assembly line put in place by management and that they argued the UAW failed to contest.

As they wrote, "A walkout occurred at the Hamtramck Assembly Plant which stemmed from a gradual speedup of the production line. The facts show production soared from 49 units to 58 units an hour, within the short period of a week. The mobility of the worker was retarded to the extent that it was difficult to keep pace."[25] The League referred to speedups that unequally affected Black workers as *niggermation,* a purposefully provocative term intended to draw attention to a specifically racialized regime of production while also grating the tongues of White liberals who disregarded that racial regime. John Watson offered the most straightforward definition in the pamphlet "To the Point of Production": "simply when you hire one black man to do the job which is previously done by two or three or four white men."[26] Mike Hamlin expanded on this definition in *Fight on to Victory: Detroit's League Speaks,* where he argued explicitly that racialization and assembly-line speedups were linked:

> Well the bourgeois response to the fact that 650,000 production workers in auto in 1947 produced 4.5 million cars and now 650,000 workers are producing 10 million cars is what? Automation. That's right. But that is not the case. I mean *no way* is automation responsible for the increase.
>
> What is responsible for that increased output is what we would call "niggermation." And what it means is that they will speed up on a particular job. If a guy can't work, or refuses to work at that rate, they fire him![27]

As discussed previously, Martin Luther King Jr. and James and Grace Lee Boggs argued primarily that automation was connected to a long history of racial capitalism in the United States that was based on an ongoing differentiation between freedom and unfreedom and employment and unemployment that drives accumulation strategies.

However, they also argued that the cybercultural revolution was a new conjuncture in which automation was leading to mass unemployment, with as yet undefined ramifications on politics, race, and the organization of the economy. The League shared with them an analysis of the cybercultural means of production that tied its techniques of extracting surplus value to the history of the plantation. Alternatively for the League, as these passages suggest, *automation* discursively obscures how the exploitation of race drives the technical composition of factory labor and surplus-value extraction.

As Jason Smith notes, when Del Harder coined the term *automation* at Ford in 1946, he was not referring to emergent cybernetic and computer control mechanisms but instead older technologies like "electro-mechanical, pneumatic, and hydraulic devices."[28] As Hamlin told *Leviathan,* it only cost Ford "$58 a unit" to build a Falcon—far cheaper than what it would cost them to actually automate a plant. Because of the ability to employ devalued Black life, "it's not worth it [to make capital investments in automated machinery]. They're holding back technological changes." The League, Hamlin said, in fact supported "full automation" because "work should be eliminated."[29] Instead of a fully automated factory, as Hamlin again argued, "what happens is, the worker can tell when they really get ridiculous because they're working so fast. And like some lines, you know, go on several floors. And sometimes a guy will be trying to run downstairs—he gets so far behind—to catch up with the cars he's supposed to be working on. So that's why those departments are so overwhelmingly black in almost every instance."[30] This offers a Chaplinesque picture of the demands placed on Black workers by automation. Here the factory is defined by spatiotemporal contradictions that require the worker to be a virtuoso, able to traverse a multilevel assembly line (similar to the assembly line in Diego Rivera's *Detroit Industry Murals,* images of which feature in the opening montage of *Finally Got the News,* although Rivera paints the workers in a far more coordinated way than the League). Yet as *drum* says in its first issue, the speed on the line is such that it also limits the worker's "mobility." Black labor power is thus called on to mediate, or even overcome, these contradictions, which in different instances requires different forms of activity and attentiveness. In analyzing the intersection of racialization and bodily expectations on the so-called automated assembly line, the League argued against a long history of not only bourgeois celebrations of technology's ability to reduce the physical demands of labor but also of a Marxist history focused on the "rationality" of factory production.

For György Lukács, in "Reification and the Consciousness of the Proletariat," in order to understand the extent to which by the 1920s the commodity form had come to structure the "total outer and inner life of society," as well as how workers become conscious of their relationship to totality and thus how to overthrow capitalism, he begins by examining the factory because it "contain[s] in concentrated form the whole structure of capitalist society."[31]

> If we follow the path taken by labour in its development from the handicrafts via cooperation and manufacture to machine industry we can see a continuous trend towards greater rationalization, the progressive elimination of the qualitative, human and individual attributes of the worker. On the one hand, the process of labour is progressively broken down into abstract, rational, specialized operations so that the worker loses contact with the finished product and his work is reduced to the mechanical repetition of a specialized set of actions. On the other hand, the period of time necessary for work to be accomplished (which forms the basis of rational calculation) is converted, as mechanization and rationalization are intensified, from a merely empirical average figure to an objectively calculable work-stint that confronts the worker as a fixed and established reality.[32]

I draw attention to Lukács here in part because the translation of *History and Class Consciousness* into English is coterminous with the League's own work, and, alongside the work of Boggs, Braverman, Marcuse, and Marx, it was central to how factory automation was theorized by Marxists in the cybercultural era. I have already discussed how for James Boggs and Harry Braverman factory automation was above all a political form aimed at reducing workers' ability to control production. Like Braverman especially, Lukács focuses in the opening pages of "Reification" on the ways that factory mechanization had destroyed any "organic" "unity" to production and in the process alienating workers from not only the objects they produce but also from themselves, becoming little more than potential "sources of error." Lukács's condensed description of the rationality of the mechanized factory is also useful for thinking alongside the League because in these pages, Lukács emphasizes in his reading of Marx (and Plekhanov) that through the rationalization of the work process and thus the fragmentation of both that process and the subject of that process (here the worker), time loses its transformational, qualitative character and becomes, "in short," space.[33]

The League's own theorization shares much with Lukács. The rationality of the automated factory requires that workers become alienated from the total work process through fragmented, repetitious, intensified labor. Yet as Hamlin shows, from the perspective of the Black worker, they do not simply become "linkages" in a "virtuoso" automated machine, as Marx put it in his celebrated "Fragment on Machines" from the *Grundrisse.* The Black worker is in part reduced to a potential source of error in the continuous flow of the automated labor process; but the worker is also required to have the capacity to not simply keep up with the machine but to make the machine work by doing the labor previously required of multiple White workers and opening one's self to bodily harm in the process, which is why "those departments are so overwhelmingly black in almost every instance."[34]

In the last chapter, I discussed how Paule Marshall's description of the continuing time of the plantation economy in Bournehills paralleled Louis Chude-Sokei's recent exploration of how modernity has called on Black life to bear the contradictions of, and White anxieties about, industrial machine life. According to Chude-Sokei, Black people have had to do the labor of making technocapitalism swing in economic and aesthetic senses, harmonizing its contradictions. As he notes, in the early twentieth century, this contradiction was named by the binary *primitive modernism.* The League's theorization of so-called automation confirms this demand, and in its work, the binary takes the form of *Black worker automation* whereby the Black worker not only is subject to the physical risks of the factory but also is required to overcome the contradictions of labor power's use in automation. This also adds a different dimension to Lukács's argument that space becomes the dominant form of the factory(-society), in the sense that the factory becomes spatialized especially as the plantation. Viewed from the perspective of the automated plantation, history and time cease to become terms for qualitative change. Instead, like in Marshall, they are repetitions of a racialized rationality of the historical present that dehumanizes the workers who not only make the factory-plantation work but also, the League argues, are central to the entire valorization process of capitalism.

Nikhil Pal Singh and others argue that discussions of capitalist development that posit slavery as prior to and apart from some "true" capitalism, as well as its normative claims about mutually contracted free labor, elide the extent to which capitalism has always been, and continues to be, reliant on the differentiation between free and unfree labor. For the League, an explicitly provocative, and to many at the time offensive, concept, *niggermation,* thus did important conceptual work. It drew attention to the physical demands placed on Black workers by the automated assembly line that largely repeated the most demanding aspects of plantation slavery, but it was also a concept that congealed an historical perspective that argued that the production and exploitation of racial difference, and the capacities of Black workers, were necessary parts of the full development of fixed capital as it manifested itself in the automated factory-plantation.

In Marx's "Fragment," "fixed capital," or large industry and ultimately automated machines, is representative of the "full development of capital" in which all of society's scientific knowledge and command of nature are mobilized as "power over living labor" and the "conquest of the production process."[35] At this point, surplus-value extraction depends less on labor time spent in production but instead depends "on the general state of science and on the progress of technology, or the application of this science to production."[36] The League departed from the Marx of the "Fragment" to the extent that instead of understanding automation as that which reduced necessary labor time to a minimum (even while it continued to posit labor time as the measure and source of value), they believed that all value continued, as Cockrel told *Leviathan,* to "flow from production"—that is, from Black labor power.[37] However, the League shared with Marx, as well as Lukács and Italian *operaismo,* a belief that the (automated) factory was at the center of capitalism. Contending that "we're not stupid people,"[38] in a long passage that is almost identical to his discussion of ownership in *Finally Got the News,* Cockrel argues that from the perspective of production and Black labor, the systemic, planetary, and contradictory features of capitalism's command over nature and labor should be seen as represented in the racialized organization of the automated, automobile factory; but he also argues that the factory is the form that organizes the planetary system of value extraction. Further, according to Cockrel, knowledge of these processes is what makes capitalism vulnerable to revolution, and the Black worker's position within that nexus of extraction and production is vital to the revolutionary process. The League's project was both to understand planetary capitalism as a whole and to produce a revolutionary consciousness in Black workers and the community equal to the task of challenging it. In this respect, they depart from the main currents of Black Power thought in general, and more specifically from Black Power cybercultural thought, by positing the centrality of the Black worker on the (so-called) automated assembly line to revolution and not the outsider or the lumpen, as others did.

Finally Got the News

In histories of the League, whether those produced internally, as in the document *Brief History of the League,* or those about them, like *Detroit: I Do Mind Dying,* the League's beginning is attributed to the intersection of two different events: the founding of DRUM at Chrysler's Hamtramck plant and the founding of *Inner-City Voice* by John Watson and others (and their subsequent control of the Wayne State student paper, *The South End*).[39] Both of these were in response to the 1967 rebellion as well as to the long-term structural contradictions of industrial, racial capitalism based on the superexploitation of Black workers, with the ultimate aim being to organize those workers in order to take state power. Unified for a time by that project, the League theorized both of these trajectories in terms of the point of production. For example, in Watson's interview "To the Point of Production," originally with *Fifth Estate* and also published as a pamphlet, he begins by discussing DRUM and the ways that surplus-value extraction in the automobile industry is based on racialized forms of labor. He then shifts to discussing how "the production of publications" is required for organizing both inside and outside the factory, as well as to supplement factory workers' technical abilities and their time. Watson discusses in detail the requirements of publishing, especially forms of expertise and the necessary machinery. In their theorization of automation, Black workers were required to overcome the contradictions of the industrial labor process that mobilized the entire scientific history of racial capitalism against them, while producing the news required a countermobilization of technical and scientific knowledge.[40]

Watson's description of cultural production in "To the Point of Production" complements Ken Cockrel's discussion of planetary capitalism in "Our Thing Is DRUM!" There, knowledge of the way that "ownership" (by "motherfuckers who don't do nothing") of nature—through mineral wealth and the means of extraction and production, as well as of people through multiple forms of debt like medical bills and consumer credit—is secured through techniques of power.[41] According to Cockrel, these include both the state's monopoly on violence but also the "University of Chicago School of Economic Bullshit," which provides the normative forms of knowledge that justify ownership's violence. All of this "flows from production"—both the production of wealth and knowledge of that system.[42] As this example shows, the League shares with Lukács not only a theory of the automated factory but also a belief that proletarian knowledge derives from its confrontation with, and necessary drive to overcome, the mediations of reified capitalism in its struggle for totality and revolution.

Alongside Black worker self-activity against the demands of automation, for the League, political struggle in the cybercultural era thus required an explicit engagement with mediation through media itself. In "Black Editor," Watson draws on Lenin to argue that the newspaper can be a form of "permanent organization, it could provide a bridge between the peaks of activity. It creates an organization and organizes the division of labor among revolutionaries. Revolutionaries do something, not just a meeting on Sunday, making speeches and passing resolutions. It creates the kind of division of labor needed not just for the newspaper but for a revolutionary organization."[43] He describes some of the activities that the revolutionary workers associated with the *Inner-City Voice* were engaged in: opening a coffee house, teaching Black history and Marxist–Leninist theory, and helping to produce other publications like *drum* and *Black Student Voice.* This suggests that in the "conglomeration of activities" around *Inner-City Voice,* a Black editor is attentive to the differences, and different needs, of these new Black organizations.

Yet despite the interview's title, the specificities of *Black* editing in this conjuncture are never explicitly stated. However, an analysis of editing in *Finally Got the News,* as well as issues of *The South End* from 1969 along with *Inner-City Voice* and *Black Student Voice,* demonstrates that the practice of editing itself was central to the League's expanded definition of production. Through editing, the League theorized a history of racial and class composition and attempted to traverse the boundary between factory-plantation production and the ways that race was put to work to (re)produce capitalist social relations in the community, especially the education system.

Finally Got the News, which was made in association with members of the Newsreel collective, was primarily an organizational film about DRUM's failed attempt get Ron March elected as a UAW representative from Dodge's Hamtramck factory.[44] It is also a document meant to spread the League's work beyond Detroit, nationally and internationally, and to show how Black workers were exploited through the collaboration of management and the union—or, as they chant in a scene on the picket line, viewers and workers watching the film "finally got the news about how our dues are being used." Alongside this, it is also the work that distills most explicitly the League's thought, especially in its use of montage to theorize the history of race's productivity for capitalism as it congealed in late 1960s Detroit.

This is especially true of the film's approximately four-minute opening montage of historical footage that was edited by Newsreel's Stewart Bird along with John Watson.[45] The first images are stills of slave auctions and an advertisement for "Negroes for Sale," slaves working in cotton fields (or at least Black tenant farmers—the images are undated), and a particularly striking nineteenth-century photograph of an older Black woman formally posed with a White baby. These initial images are scanned by the camera, either moving up and down or back and forth, as if the camera, while explicitly positing their importance, is also searching for their historical meaning. The cuts in the opening minute or so are relatively slow, and the images are accompanied by persistent drumming that is both a clear reference to DRUM and an evocation of the continuities between American and African music and the importance of Black music in Detroit. Between the first and second minute, while the drumming continues, becoming quieter and then building in volume and speed, the images shift from scenes of slavery to industrial ones, most often pictures of White workers in the factory or on strike. Here again, the camera scans the still images, highlighting different aspects of each, the back-and-forth of the camera often leading into the next image and the camera movement helping the viewer transition between often rapid cuts. At about two and a half minutes, the drumming gives way to funk and rhythm and blues, the speed of the music increases, and the montage is intermixed with pictures of White suffragettes and labor activists and, increasingly, images from the Civil Rights Movement and the Detroit uprising of 1967. At the four-minute mark, the credits roll, and John Watson begins a living room lecture on race, surplus-value extraction, and the revolutionary potential of Black workers.

In her analysis of *Finally Got the News* as a document of the transition between Fordism and post-Fordism, Morgan Adamson utilizes Ian Baucom's concept of the "enduring image," which moves into the future by doubling back on itself, to argue that the film's opening montage connects the uprising of 1967 to historical struggles dating to plantation slavery: "An enduring image of the American century is the image of Detroit; specifically, it is the image of struggle that repeats and intensifies at the end of the period. *Finally Got the News* is this very image."[46] I agree with Adamson's assessment that not only is *Finally Got the News* one of the key theoretical texts of the period, but that in many respects the opening editorial moment does as much theoretical work as subsequent lectures by Watson and Cockrel. This is especially true in that it suggests that race is (re)productive for capitalism in ways that are both historically continuous and also, and necessarily, discontinuous in terms of the relationship between past and present as well as in terms of the ways that race, class, and gender overlap but never correspond.[47]

The montage shares the exploration of this contradiction with the beginning of the internal theoretical document "The History and Derivation of the League of Revolutionary Black Workers." The beginning of the text is in dialogue with Karl Marx and Theodore Allen, whose "Can White Radicals Be Radicalized?" challenged then-mainstream Marxism's, and much of the New Left's dismissal of race in discussions of class in the United States. In it, they quote from a 1846 letter of Marx to P. V. Annenkov where he wrote that because "slavery has given their value to the colonies . . . [providing] the necessary condition of large-scale machine industry . . . slavery is thus an economic category of the highest importance."[48] Immediately following this, the League critiques Marx: "But American slavery is more than a 'mere' economic category. The *social relations* between men in American society are transformed from relations between *classes* into relations between *races*. For racial slavery represents a transformation of slavery *itself*, since it converts slaves themselves from a class of slaves into a race of slaves. American society is transformed from a class society into a 'race' society and Racism is born."[49] In *Finally Got the News*'s opening montage, the series of images of slavery is the opening moment in a longer history of the racialized (de)composition of class. Here forms of labor normatively defined as "White" in the plantation period, often factory labor, and those defined as "black," agricultural work and certain forms of domestic labor, are separated from each other and are thus implied to have different values. Notably absent from the entire montage is the Black industrial labor that is the primary focus of the film. In this respect, the montage is a prehistory of the revolutionary Black assembly-line worker. Even though it contains images of successful struggles against capitalism, such as those to form the CIO and UAW in the 1930s, these can be taken as only one side of the history of anticapitalist struggle in the United States, in the sense that the history of White working-class struggle did not prevent the conditions of possibility for the political-economic crisis of 1967. The opening montage, while largely triumphant, can be read as making an argument for the continuity of racialized labor and racialized struggles for freedom leading up to the Detroit rebellion, precisely because, as they write in "The History and Derivation of the League of Revolutionary Black Workers," "white skin privilege" prevents actual class formation.[50]

However, the montage also tells us that a number of contradictions—here, literally questions—exist as to the relationship between, race, class, and gender and the ways those are mobilized to produce capitalist social relations. These uncertainties are opened by the two "question marks" that interrupt the photographic images. The first comes at one minute and forty seconds, in between a photograph of labor leaders and management shaking hands (an image animated by the camera itself moving up and down) and images of confrontations between police and Black agricultural workers as well as ones showing confrontations between police and White workers and White suffragettes or labor activists. The second happens at approximately two minutes and thirty seconds, where the insertion of a question mark comes between images of strikes and the moment when contemporary funk, moving images of the factory, and images of the 1967 Detroit rebellion replace drumming and historical footage. On the most obvious level, given that the question marks follow images of union and corporate agreement, these can be read as critiques of the UAW's relationship with management. However, the subsequent images are of Black workers and the political activism of White women in the early twentieth century. The question mark here literally does the work of questioning. What is the relationship between labor contracts, police violence, race, gender, and political activism? Such questions cannot easily be answered through linear narratives or direct connections, given that capitalism relies on exploiting the contradictions inherent in these categories of social difference.

In "The History and Derivation of the League of Revolutionary Black Workers," a particularly interesting moment follows their expansion of Marx's sense of the relationship between race and class. They write that because of racial difference, it is not simply that a normative version of the "working class" is never produced, but that "white wage labor" and "Black labor"—the former because of "white skin privilege" and the latter because of the legal restrictions of (post)slavery—are never "fully proletarianized," with *proletarianization* defined here, quoting Marx, as being "stripped of all qualities except that of being labor."[51] I have more to say about proletarianization in the next chapter; however, here this emphasizes that these categories remain productive for capitalism because even in the historical struggles shown in the montage, they are never emptied of their so-called natural qualities, whether this be, for example, Whiteness or masculinity, which would expose them as social relations of hierarchical difference that are exploited to divide potential allies. Thus, the montage, by its questioning, suggests that even in racial and gendered political alliances, something remains that prevents those subject positions from being emptied as the first step toward entirely different social relations, the unfinished historical project that *Finally Got the News* goes on to argue the League was involved in.

Although ostensibly different, *Finally Got the News*'s opening montage of the history of Black and working-class struggle against capitalism leading up to the Detroit rebellion and the League's formation parallels its theory of Black labor on the automated assembly line. Far from marking an entirely new moment when automated unemployment produces a new composition of race and class, race remains productive for both the intensified labor that makes "automation" work and productive for divisions within class struggle that capitalism is able to exploit in the extraction of surplus value. In their theorization of the assembly line, they make direct connections between Black labor and automation. In *Finally Got the News,* montage supplements these direct connections and allows them to juxtapose seemingly disconnected moments in the production of social difference in the world beyond the factory as well as in it, like social solidarities that are never fully emptied of their particularities. In this sense, *Finally Got the News* is an example of *Black* editing, even though it is a collaboration between Watson and White filmmakers, because it foregrounds the production of racial difference across boundaries in the history of Black working-class struggle for state power.

In "To the Point of Production" and elsewhere, Watson argues that film is an especially important "vehicle for the reeducation of black people to the true nature of the system."[52] The choice of words here is felicitous: it connects the production of automobiles by Black workers to their production of systemic knowledge in the medium of film. Jonathan Beller argues in *The Cinematic Mode of Production* that cinematic "montage extended the logic of the assembly-line (the sequencing of discreet, programmatic machine-orchestrated human operations) to the sensorium."[53] For Beller, *cinema* names more than movies and instead names an organization of life such that looking becomes value productive for capitalism.[54] In Watson and the League's thought, film was especially important because of its ability to reach a global audience. However, film production did not stand alone, and the production of the news in various media, especially print, was always edited together. In *Finally Got the News,* the cinematic "production of the meaning," as Beller calls it, not only of how racial capitalism is organized but also how the League was a counterorganization against its logics, comes midway through the film.[55] In the middle of discussions about Ron March's campaign for president of UAW Local 3, there is an extended montage of people producing various League publications. Scenes of reading, writing, editing, and clipping, as well as the movement of the printing press, are intercut with images of election pamphlets, issues of *elrum,* Arabic-language materials, and the cover of the April 1, 1970, issue (volume 2, no. 3) of *Inner-City Voice,* which featured the DRUM slate on its cover. The montage produces not only a way of seeing the election but also a way of seeing the League as a media, in the sense that an intermedia apparatus of film and print is deployed, and in doing so mediates both diverse Black organizations and knowledge of how racial capitalism works historically as well as in then-contemporary Detroit.

This intermedia (or cinematic in Beller's sense) production of knowledge, although most explicit in *Finally Got the News,* is also evidenced in their print productions. Issues of *The South End* from 1969 are especially representative of this. For example, the January 23 issue (volume 27, no. 62) leads with the headline "DRUM—Vanguard of the Black Revolution/Dodge Revolutionary Movement States History, Purpose and Aims." The issue's second part is entitled "Stage II: Building the Organizational Structure" and includes images of non-DRUM revolutionary figures like Huey Newton as well as local advertisements and a classified section. This issue is less a traditional newspaper with discrete articles than a series of subsections on the founding, activities, and organization of DRUM. It opens with a "History" that announces DRUM's revolutionary character; it also notes that "the complete and total transformation of the society" will require more than just factory workers; it will take "the whole black community as well as other progressive sectors of the rest of society." Here the editing together of factory workers, the Black community, and progressives is done through the work of feature editor Luke Tripp.[56] The issue lacks any attributed authorship; instead, newly written material is intercut with quotations, both cited and uncited, from issues of *drum.* For example, "History" begins with the newly written materials I just quoted. This is followed by a clearly marked citation of *drum,* volume 1, no. 13, disclaiming any responsibility for racial divisions during the first wildcat strikes. Here the citation of *drum,* volume 1, no. 13, appears in bold print; DRUM's new commentary on its original piece, along with further critiques of the UAW, is in lighter print. However, the citational practices are far from consistent, and the boundaries blur between newly written and quoted work, creating the sense that what is produced in the factory and what is produced outside the factory are themselves blurred.

These editing practices are evident in other issues as well. The April 30, 1969, issue (volume 27, no. 113) is primarily a long multipart article on economic depression entitled "The Needy and the Greedy." It argues that the "U.S. economy is currently in a thinly disguised state of collapse" and situates the dispossession of Black labor in relation to "over-production" and continued White capitalist wealth.[57] In ending, it includes an unattributed reprint of *drum,* volume 1, no. 14, on "Black Power" that argues capitalist wealth is rooted in Black labor power, and it is only through collective action against capitalism that depression can be avoided. The intermedia editing that appears in *The South End* not only moves between it and *drum,* but also between *The South End* and other publications. For example, the May 22, 1969, issue (volume 27, no. 126) contains a long article, "The Midwest," about how the reorganization of capital and labor in Detroit is connected to the regional economy. This article is then used as the beginning of the long *Leviathan* article "Our Thing Is DRUM!" that I have already drawn extensively on. These are only a few of the many examples in which news production in the League was defined by editing practices that theorized the historically contingent relationship between race, class, and the organization of the labor, but that also mediated the factory and the community. These practices flow from an expanded sense of production whose techniques and boundaries the League openly theorized and experimented with. Yet if editing allowed the questioning and juxtaposition of that which in then-normative U.S. Marxist thought seemed unrelated, like race and class or factory and community, the extent to which the League grappled with the broader politics of social reproduction remains in question.

Poetics

Dan Georgakas, in his sympathetic critique of *Finally Got the News,* wrote in *Cinéaste,* "The second half of the film never regains the sharp ideological and artistic focus which ends with the Dodge election. There is a valiant attempt to deal with the relationship of white workers that might have been an entire second film. . . . Similar problems surround the treatment of black working women and the role of community organizing. Hurried sequences do little more than register the film's awareness that such topics need further analysis."[58] These scenes primarily focus on White workers formerly from Appalachia, reflecting on the limitations of factory work and life in Detroit, supplemented by Cockrel and Watson discussing the extent to which whiteness limits White workers' ability to identify themselves as not simply in need of more money and better working conditions but as exploited by capitalism in ways comparable to Black workers. These scenes are thus continuous in many respects with the League's theorization of race and class and the incompleteness of proletarianization in the film's opening montage and in other media. While the discussion of White factory workers involves multiple interviews and extensive theorization by Watson, the scene of Black women's double oppression is even more limited. Lasting only approximately two minutes, it is narrated by an uncredited woman and discusses the police killing of nine-year-old Danny Smith as well as Black women's limited employment opportunities.[59] Both of these passages follow the section on news production that I analyze above, and in this respect, they mark the limits of the League's sense of production and their media practice of editing. Despite claims by Cockrel in "Our Thing Is DRUM!" that capitalism is "cannibalistic," feeding off the life-making capacities of working-class people, neither these scenes nor their work in general ever fully coheres into a theory of how race and gender are (re)productive for the extraction of surplus value in the factory and into the community at large.[60]

The stated goal of the LRBW, as well as the many revolutionary unions, was to first gain control over production in automobile assembly lines and especially to reduce the intensity of production, and then to expand their power base beyond the point of production, arguing that it is only by taking state power that the means and relations of production could be changed in order to end capitalism and create an economy based on a just means of distributing society's surplus.[61] This program "flows from the point of production" and the political theory that Black industrial workers are the linchpin of U.S. capitalism. The limits of the point of production as it relates to both a theory of capitalism and the way it uses race and gender in the organization of value is evident in those moments where Black women are addressed.[62] However, their discussion of education is one moment in which their theory of production becomes a theory of the social reproduction of racial capitalist relations that expands beyond Georgakas's apt critique of *Finally Got the News,* even if terminologically they stay within "production."

In his speech "On Repression," Ken Cockrel defines the education system in terms of the "productive relations of society." Using the example of Northwestern High School (now Detroit Collegiate Preparatory Academy at Northwestern), he argues that schools do not simply educate Black students for assembly-line and menial labor; rather, they are "deliberately structured in such a way as to produce potential unemployed men during periods of what are euphemistically referred to as recessions," keeping in mind that at this moment, the League was arguing, much like Martin Luther King Jr., that U.S. capitalism was on the verge of entering a depression that was already manifesting itself in Black communities.[63] Schools in Cockrel's formulation were akin to Louis Althusser's ideological state apparatuses, in which the relations of production were reproduced, even if Detroit's schools were failing to interpellate Black students into the racial capitalist regime of accumulation because of the Black Power and civil rights movements. Elsewhere, Cockrel refers to "apparatuses of state power" in ways that bridge Althusser's ideological and repressive forms.[64] Notably, unlike other Black Power cybercultural theorists and the period's automation discourse generally, automation does not lead to a society of mass unemployment because automation, strictly defined, doesn't exist. Intensified Black labor and a reserve army of labor are both products of the long history of capitalist accumulation strategies, not technological development. By extension, schools are not outdated forms that fail to produce students for some new cybernated economy. Instead, they continue to do exactly what they are designed to do: they are sites of struggle over the reproduction of the relations of production, which always requires a surplus Black labor population.

As Cockrel notes, the only way to have a society not for profit "but production for the use of those who do the producing" is to take state power. However, in "Our Thing Is DRUM!" he emphasizes that their program for leading struggles in the factory, schools, and in the community is "based on reality" and not the "metaphysical."[65] This echoes the Boggses' Marxist sense of scientific Black Power, which tried to organize a classless society out of the historical contradictions of actually existing cybercultural capitalism. One of the most important aspects of the League's organizing was the formation of *Black Student Voice,* a newspaper produced by Black students at Detroit's Central High School.[66] Like *The South End, Black Student Voice* sought to contest from within the educational system the extent to which it was what Luke Tripp calls "the ideological and technological servant of the American ruling class."[67]

The first issue is especially representative of this project. Including articles that might fit in any student paper—for example, an article about the misuse of student council funds—it also includes a "Tribute to James Johnson," a cause célèbre of the League who was acquitted of a factory shooting under the strain of assembly-line speedups. Most prominently, it features a comic about "James Anthony," a student called out for laziness by his teacher, especially because his high standardized test scores rank in the top 1 percent nationally.[68] This leads James to respond that he does not pay attention in class because he goes to a school that is rat and cockroach infested, where the playground is covered in broken glass, and where the "bullshit curricula [are] designed to systematically rob minority youth and prevent [them from] acquiring of any skills what-so-ever." After being admonished by the principal for not appreciating that he's being educated to get a factory job making "$3.50hr," with other students he forms a newspaper as the first step toward organizing the "Black Student United Front." Critiques of the League often point to its top-down structure; however, *Black Student Voice* is an example of how the forms of popular cultural productions were *détourned* by students themselves to show how the educational system operated as an institution that attempted to reproduce the dominant composition of race and class—and this at a moment when Fordist–Keynesianism was increasingly in crisis.

The League's cultural productions, like *Black Student Voice, Finally Got the News, The South End,* and *Inner-City Voice,* sought to organize Black workers and the community; it also sought to produce forms of revolutionary culture and education that would complement the struggle over the point of production. In the internal document "Revolutionary Black Culture as Weapon," they examined the effectiveness of historical forms of Black revolutionary culture, especially the United Negro Improvement Association, as well as contemporary examples that had "seized the time" in creating a new Black culture, like the Nation of Islam, the Black Panther Party, and the organizations associated with Amiri Baraka, alongside the examples of the Chinese and Vietnamese communist parties.[69] They primarily argue here that, unlike cultural nationalist organizations, they needed to create a Black workers' culture that would include, among other examples, cultural productions like *Finally Got the News* and a series of Black worker holidays.

I end this chapter with an analysis of a prose poem by Bobby Jean Cummings, an author whom "Revolutionary Black Culture as Weapon" names as exemplary, if for reasons it never elaborates. "Poetics" was a semirecurring section in *drum* that included work by different authors, the first of which was Cummings in volume 1, no. 2. Cummings's piece is addressed to an anonymous "you," implied to be White America: "You are my Jesus," Cummings writes, suggesting that White America has created a system structured in racial inequality as the ideal. However, the results of this are that the author, as metonym for Black America, has become "your waste product . . . your Frankenstein monster. Yes Lord." Although it begins with religious metaphors, the piece is primarily defined by machinic ones; the racial monster is constructed by American technological modernity as a subservient (re)productive machine. However, by the poem's end, the self-conscious Black subject will "threaten" not only America's machines but also its racialized standard of "universal" value. As this suggests, such a racial cyborg is conscripted to do multiple types of labor: "You brought me to America in chains, I built your America. I built bridges and railroads. I composed poems. I sang lamentations. I nursed your children at my breasts. All this I gave free of charge, it was all yours for the asking. Yes, Lord." The multivalent meaning of the title "Poetics" is brought out in this passage. Forced into slavery, the Black Frankenstein monster is not simply that which is made into waste or left over. They are Black Power in the sense defined in the pages of *drum* and other League texts: Black labor power, which built the country's infrastructure. At the same time, they are also responsible for making America in a broader sense that includes the country's culture, while also doing the work of social reproduction by giving life to both White and Black Americans, echoing the opening montage of *Finally Got the News,* which shows Black workers in the fields and as domestic laborers raising White children. Notably, in this piece there are no gendered pronouns, and it is one of the few moments in *drum* that is not written from a masculine perspective. This non-binary presentation thus allows the work to move between the productive and the reproductive and between the spaces of machine and culture, which is unique in the League's work.

In the first half of the poem, these activities are, for the narrator, part of the striving for whiteness: "You warned me to strive for the best, to be you. You said that white is the standard, the universal, black, the particular. According to you the particular has no meaning, no significance, no existence of its own. It is an imperfect copy of the real, the universal. You told me that the universal is white skin, straight hair, thin lips, sharp nose. Yes Lord. . . . Oh, how I believed you." The influence of Hegel and Frantz Fanon are stronger here than in any other moment in the League's work.[70] Identity is mediated by racial otherness in terms of the relationship between a given subject racialized as "black" (and particular/inferior) to one racialized as "White" (and universal/the standard), but also in the sense that the White universal also mediates relations between Black peoples. With the categories of *universal* and *particular,* the register here is philosophical, but the poem is also embedded in the specifics of technocapitalism in the 1960s. The author attempts to become White by "merg[ing] with the god head." "In the process," the author writes, "I [relinquish] my consciousness, became a program machine, a non-thinking being. Well, now I have no need for gods: I don't need you because I have discovered myself, my potential. I can become anything that I wish." As I discussed in the last chapter, Allen Fuso defined himself as an IBM machine incapable of sexual, domestic, or embodied social relations, problematically in opposition to Black life in Bournehills. This formulation was part of a technocultural moment in which, as Fred Turner writes, "the brain [was imagined as] a form of digital hardware and its actions a form of software, that thinking was a type of computing and memory simply a matter of data retrieval."[71] Paule Marshall, Eldridge Cleaver, Alexander Weheliye, and Louis Chude-Sokei, among others, argue that the disconnect between the mind and body in post–World War II cybernetic thought was able to emerge because of a racialized history in which whiteness is associated with a disembodied intellectualism and blackness with the body. Here, however, this binary is traversed. While the Black subject of the poem is responsible for the making of America industrially and reproductively, technocapitalist universality seems to not lead to the computer, or "program machine," as a form of disembodied mind; instead, the program machine is an embodied, nonthinking form that only repeats a history of racial inequality.

In becoming a program machine in the cybercultural era, a moment of crisis in the technocapitalist mode of production that required, according to the League, intensified Black labor instead of mass automated unemployment, the narrator undergoes a transformation, writing, "I have destroyed your creation, I have created a new being." How this moment of poetics happens is unclear; there is no explicit process of Hegelian struggle through which the enslaved person becomes free, although this is strongly implied. From this point, *Black* takes on a new meaning; it is no longer a mark of inferiority but a positive sign of "existence." Much like the "ontological totality" of the Black Radical Tradition in Cedric Robinson's *Black Marxism*, the subject now realizes that "I existed before you came to Africa," that "I am a giant shadow, a shadow lurking in the background and threatening your whole existence: your institutions, your buildings, your machines." It ends, "Be assured that I don't know who you are: I only know that I am, that I exist."[72]

Bobby Jean Cummings's "Poetics" and the totality of the LRBW project should not be reduced to each other. However, "Poetics" distills in many respects the League's theoretical attempts to theorize capitalism from the perspective of Black labor power at the point of production. The history of the United States flows from racialized production that makes capitalism possible. While "Poetics" is more effective than most of the League's work at describing how race was both productive and reproductive for capitalist social relations—a necessary category for bringing the system into existence and managing its contradictions—it embodies the League's intention to create a Black worker and community consciousness aware of its "essential and key" position in the cybercultural capitalist machine, and thus its capacity to shut down and ultimately transform that system.

6

The Black Panthers

"PEOPLE'S COMMUNITY CONTROL OF MODERN TECHNOLOGY"

In the June 26, 1971, issue of *The Black Panther: Intercommunal News Service,* the party announced that "Like All Things," the party "Does Not Stand Outside of Dialectics." Because the Ten Point Program from 1966 no longer fully represented either the conjuncture of 1971 or the "desires of our people," the article announced that it would no longer be published in the paper, as it had been since the second issue on May 15, 1967, until it "more accurately reflects and defines the present needs and desires of our people."[1] The new Ten Point Program would not be published until May 13, 1972, after having been presented publicly at the Community Survival Conference in Oakland on March 29.

The original program was drafted by Huey P. Newton and Bobby Seale on October 15, 1966, and was, in part, a *détournement* of the U.S. Constitution and Declaration of Independence. As Robyn Spencer has described it, it was a "program of radical reform rooted in the nexus between visceral lived experiences of discrimination in Oakland's flatland and the strong sense of connection with forces of radical social change around the country and the world."[2] Between the original version and the 1972 revision, there are a number of significant changes. For example, in the original, point 1 reads, "We Want Freedom. We Want Power to Determine the Destiny of Our Black Community," which was revised to "We Want Freedom. We Want Power to Determine the Destiny of Our Black and Oppressed Communities."[3] This change is representative of the document's expansion to include other oppressed communities alongside the Black community—a change closely tied to Newton's theory of intercommunalism, which by then had become the guiding ideology of the party. Other significant changes included more explicit references to capitalist, and not just racial, exploitation, as well as an entirely new point 6 demanding "completely" free health care for Black and oppressed communities instead of an exemption of Black men from military service.

In this chapter, I focus primarily on another aspect of the revised program reflective of the theory of intercommunalism. Point 2 continued to demand full employment in the 1972 version; however, it was expanded to include not just control of the "means of production" if full employment or a guaranteed income was not provided by "American businessmen," but also to include community control of "technology." This was consistent with a change to the program that entirely dropped point 10's demand for a U.N. plebiscite on national self-determination (which was not included in the first 1967 printing) to expand the demand for "Land, Bread, Housing, Education, Clothing, Justice, and Peace" to include "People's Community Control of Modern Technology."

What is at stake in the inclusion of "People's Community Control of Modern Technology" in the Black Panther Party's Ten Point Program? To ask this is to raise what Huey P. Newton calls in 1972 "The Technology Question."[4] There, Newton argues that the technology question revolves around two issues. First, technology is a capitalist (neo)colonial organization of life such that a "reservoir of information," a form of primitive accumulation for the cybercultural era, was produced and appropriated by, primarily, a U.S. capitalist "ruling circle" in order to create the world of "Reactionary Intercommunalism" or "empire," as well as new forms of consumerism. Second, he argued that oppressed communities required a "worldview" attuned to the new planetary organization of technocapitalism in order to contest it. In concluding *On the Eve of the Cybercultural Revolution,* I argue that the technology question, as Newton formulated it, was central to the Black Panthers' politics. Approaching the Panthers through the technology question demonstrates the extent to which they were capacious theorizers of the politics of race, class, technology, and capitalism in the cybercultural era, as well as to the extent to which their thought acknowledged that the world was constantly changing and was thus open to horizons of the future that ask us to reconsider the political concepts that we bring to our own analyses of technocapitalism in the early twenty-first century.

Through an examination of often neglected essays, like "The Technology Question" or Elaine Brown and Angela Davis's theoretically rich discussion "Welcome Home, Angela Davis," this chapter focuses on two specific aspects of their technopolitics. The first section examines the role of the lumpenproletariat in their analysis of the cybercultural revolution, and the second looks at the 1970 Revolutionary People's Constitutional Convention (RPCC) in relation to the Black Panthers' survival programs and Huey Newton's theory of intercommunalism. The Panthers' politics of technology is, however, much more extensive than these two aspects and would require more than a single chapter to examine in depth. The first appearance of technology in the Black Panthers is undoubtedly the guiding quotation of the first issues of *The Black Panther: Black Community News Service:* Huey Newton's "The Spirit of the People Is Greater than the Man's Technology." Here, it is entirely decontextualized; however, its origin lies in the first action that brought the Panthers to the attention of the Black community in Oakland. Denzil Dowell was killed by police on April 1, 1967, in Richmond, California, and, unable to get answers or redress from government officials, his family turned to the nascent Black Panthers for help.[5] As Bobby Seale describes it in *Seize the Time,* while Newton was speaking at a rally for Dowell, a helicopter "kept flapping over and Huey pointed up at the helicopter as it was going over and said, 'Always remember that the spirit of the people is greater than the man's technology.' And the people said, 'Right on.'"[6] Technology is here defined, as it often was, as the machines of the military, policing, and surveillance apparatuses that the U.S. government would bring to bear in its "war against the Panthers," as Newton called it in his dissertation at U.C. Santa Cruz.[7] Perhaps the most striking example of this conception of technology is an article in the *Black Panther* from May 18, 1968, entitled "Computers to Watch Riots" about two new computers that the federal government had recently acquired that would allow the FBI to collect and process information about urban rebellions in order to more effectively deploy new antiriot weapons like lasers, Teflon, foam, and cattle prods.[8] The Black Panthers' political program, and one of the most explicit reasons why people's community control of modern technology was necessary, was to counter government repression, and this article is accompanied by a two-paneled Emory Douglas illustration showing "Pigs" in a state of alarm as guerrilla warfare overwhelms and scrambles the FBI computers.

These brief examples give insight into the context of the Panthers' theorization of the politics of technology. In this chapter, however, I especially hope to show them at their most speculative while also showing that, especially, Newton, Davis, Brown, Eldridge Cleaver, and George Jackson did so with the aim of producing useful political knowledge from the perspective of Black, working, and oppressed communities. As Newton speculated in the 1974 essay "Mind Is Flesh," the most likely outcome of the reign of Western philosophy's conception of mind–body dualism, as it was reorganized by the politics of the COINTELPRO era, was the implantation of electronic devices in human brains in order to surveil and control the population.[9] The political task of this mode of "dialectical materialist" theorizing was not speculation for its own sake but "to invoke the entire human body in a complete social-historical context" in order to show not just how scientific conceptions of what constitutes the human have been constructed through racial, colonial, and class difference but also how knowledge of this history can aid oppressed peoples in creating a "posthuman" "future era of abundance" that is based on alternative uses of those technologies.[10]

It is in the spirt of Newton's dialectical materialist philosophy, which "employ[s] a framework of thinking that can put us in touch with the process of change," that I approach their politics of technology.[11] The first section examines the "confusion," as Eldridge Cleaver called it, surrounding the place of the lumpenproletariat in the Panthers' ideology. Marx and Engels famously contended in *The Communist Manifesto* that lumpenproletariat, "the social scum, that passively rotting mass thrown off by the lowest layers of the old society, may, here and there, be swept into the movement by a proletarian revolution; its conditions of life, however, prepare it far more for the part of a bribed tool of reactionary intrigue."[12] Alternatively, influenced by Frantz Fanon, the Panthers placed, for a time, the Black lumpenproletariat at the center of their revolutionary politics. In critical examinations of the Panthers, the lumpen feature prominently and controversially, yet their role, and why it was contested by other figures, was tied to their analysis of the composition of race and class in cybercultural capitalism.

Cedric Johnson and Jackie Wang are among the few to have considered the Black Panthers as theorists of technocapitalism. Wang argues that the most significant aspect of the Panthers' conception of the lumpenproletariat is that "labor-saving technologies will not necessarily liberate humans from work" but that they can "lead to the creation of surplus populations that are housed—and generate value—in prison or are folded into the economy as debtors."[13] My own approach will sidestep evaluations of the political efficacy of the lumpenproletariat, as others have done, and instead follow Wang in theorizing the lumpen as a concept that helps us to understand the contingent relationship between racialization, technology change, and politics in the cybercultural era.[14] However, unlike Wang, an examination of their work shows that *lumpen* was in fact a contested concept that, although important, was less prominent in their political theory than it is often assumed. I stage a conversation between Cleaver, Brown, Davis, Jackson, and Newton over the concept of the lumpen that in life was undermined by incarceration, government repression, exile, death, and expulsion, and that moves from discussions of the lumpen to that of neoslavery, the worker, and ultimately the community.

Huey Newton's theorization of the community, intercommunalism, and empire serves as a point of articulation in the chapter. In Newton's formulation, the Panthers' survival programs were a way to produce a new worldview and a consciousness of how reactionary intercommunalism and technocapitalism, at the end of Keynesianism, organized the world as metabolic rift and into fragmented, oppressed communities. I consider how, through politics, theorizing, and organizing a political version of STEM, the Panthers sought to create the means to facilitate the people's community control of modern technology. This, however, was only one aspect of the politics of intercommunalism. I end the chapter by analyzing the Panther-led Revolutionary People's Constitutional Convention, which sought to create a new constitution that countered the technocapitalist organization of life, and which put pluralism and the full political participation of the world's communities at its center.

This, in conclusion, returns to the issues discussed in chapter 1. Martin Luther King Jr. attempted to expand the welfare state in order to confront a history of racial inequality and a reorganization of Keynesian state capitalism by cybernation and automation.

Writing and organizing only a few years later, Newton, anticipating much about today's neoliberal world, believed that empire had already undermined the nation-state as the form through which a just distribution of surplus and alternative anticapitalist forms of social reproduction could be organized. In the early twenty-first century, capitalist and extractive automation technologies are deployed as a planetary formation, so the social relations of technocapitalism are likewise planetary. This raises the question of whether the nation-state or some other form is the best location from which to organize more-than-capitalist alternatives. Huey Newton would tell us that the contradictions of the present are different than those of the 1960s and 1970s. Yet I strongly believe that returning to how the Black Panthers and King, as well as James and Grace Lee Boggs, Noah Purifoy, Paule Marshall, and the LRBW, confronted these questions can be a guide to our own time.

From the Lumpenproletariat to the Community

The lumpenproletariat was one of the Black Panthers' most important political concepts. In *Seize the Time,* Bobby Seale writes that his and Newton's interest in the lumpen both preceded and was central to the party's founding. Drawing on Frantz Fanon's valorization of the lumpenproletariat in *The Wretched of the Earth,* Newton argues, according to Seale, "that if you didn't organize the lumpen proletariat, if the organization didn't relate to the lumpen proletariat and give a base for organizing the brother who's pimping, the brother who's hustling, the unemployed, the downtrodden, the brother who's robbing banks, who's not politically conscious 'that's what lumpen proletariat means' that if you didn't relate to these cats, the power structure would organize these cats against you."[15] Here, the lumpenproletariat is a potential political force, similar to but not synonymous with James Boggs's outsiders, who, disorganized by racial capitalism, are antagonistic to its bourgeois cultural mores and legal strategies of accumulation ("illegitimate capitalists," to use Newton's terminology). Consequently, they are also politically disorganized and primed for co-optation by the power structure. In this respect, they are much like the classic figure of the ragged, dispossessed, and propertyless class so disparagingly described by Marx and Engels.

As important as the lumpen was to the Black Panthers' ideology, it has played probably an even more important role in critical examinations of the Panthers, where its centrality is nearly taken for granted.[16] This is for good reason; the concept is prominent throughout the Panthers' multimedia oeuvre, as Seale's example suggests. However, as Eldridge Cleaver wrote in "On the Ideology of the Black Panther Party," "There is a lot of confusion over whether we are members of the Working Class or whether we are Lumpen-proletariat. It is necessary to confront this confusion."[17] This productive "confusion" is in fact at the heart of the lumpenproletariat as a political concept and is especially pronounced given that, as Jackie Wang argues, they mobilized it at a moment when capitalism and the carceral system were being reorganized and expanded. Yet while the category of the lumpen itself is undoubtedly of great importance, an examination finds that it is also in productive dialogue with other competing categories to describe the politics and economics of the conjuncture. This is in part because all of the Panthers I examine here are explicitly dialectical Marxist theorists whose concepts are in a state of continual development—a feature that is especially pronounced given the political-economic shifts that I've been describing throughout this book. This, however, should not be taken as a weakness. Instead, as the final example of Black Power cybercultural thought, they represent the challenge of theorizing and practicing a politics of freedom in a highly complex revolutionary technocapitalism.

Seale attributes the lumpen to Newton's pre-Panther theorizing, but it does not appear in *The Black Panther: Black Community News Service* and other public writing until summer 1970 (the same year that *Seize the Time* was published). For example, in the May 19, 1969, issue, two articles appeared that would seem ripe for some consideration of the lumpenproletariat. "Up from Capitalism," by Larry Jones, focuses on class oppression in the United States and the extent to which "class is dominated by color."[18] "Unite" by Charles Bursey argues that "we are fighting a class struggle," and that both capitalism and the struggle against it are driven by contradiction, as explained by the study of "Marxist–Leninist" works by Newton, Cleaver, and Mao.[19] Yet the lumpen is entirely absent in both of these. Although there are many potential explanations for this, including that neither found the concept particularly useful, in "Ideology" Cleaver attributes its general absence to the fact that Newton's imprisonment limited the diffusion of his ideas.[20]

The first extensive public analysis of the lumpenproletariat is Cleaver's "On the Ideology of the Black Panther Party," even if there he attributes its significance to Newton by way of Fanon and Mao, as Newton will in turn do in regards to Cleaver in a speech at Boston College in November 1970.[21] Writing from exile in Algiers, Cleaver, like the Boggses and the LRBW, frames the piece as a revision of Marxism–Leninism from a "new, strictly American" perspective, which is to say through the historical centrality of racial and Black colony antagonisms. The need to rethink Marxism for the specificities of the cybercultural United States extends to Fanon, the Panthers' most important precursor. He argues that the urban nature of the Black lumpen means that they differ from Fanon's formulation, as does their primary form of political action, the "spontaneity" of urban rebellion. Above all, theorizations of the political and economic status of the lumpenproletariat must be reevaluated to take into account the racial composition of the United States and the changing technological situation.

Cleaver makes a number of significant distinctions regarding the lumpenproletariat and the working class that are central to analyzing the technological organization of racial capitalism in the cybercultural conjuncture. First, as I already discussed in chapter 1, the United States is characterized by a system of domestic colonization that for Cleaver is largely defined by a racialized lack of ownership and control over the means of production (which, alternatively, is exactly what makes all Black people, at least in Detroit, "workers," according to Ken Cockrel). Second, there is a difference between "Mother Country," White, working class, and lumpens and "Black Colony" working class and lumpens. This might seem obvious, given that lumpens are here defined by both a lack of stable employment and by the racial cleavages of colonialism. Cleaver, however, makes another important distinction: in the mother country, the divisions between the working and lumpen classes are "fairly stable, but when we look at the Black Colony, we find that the hard and fast distinctions melt away. This is because of the leveling effect of the colonial process and the fact that all Black people are colonized, even if some of them occupy favored positions in the schemes of the Mother Country colonizing exploiters." Crucially, Cleaver argues that because of race's effect on class composition as well as on the political composition of potential anticapitalist movements, there is no "All-American Proletariat; one All-American Working Class; and one All-American Lumpenproletariat."

In "Huey P. Newton and the Last Days of the Black Colony," Cedric Johnson levels three critiques at the Black Panthers' theory of the lumpen political vanguard: (1) that guerrilla warfare led by a lumpen vanguard was not applicable in the sociopolitical and geographic context of the United States, as it was in much of the Third World; (2) it "unwittingly reinscribed prevailing Cold War ideology" by disempowering labor unions and socialist movements as political actors in the anticapitalist, antiracist struggle; (3) it used a nineteenth-century category invented by Marx and Engels that simply was not relevant in the advanced industrial economy of the late twentieth-century United States. Instead, Johnson argues that the term "industrial reserve army" would better capture the ways that waged labor and unemployment were distributed by capitalism in the United States.[22] All of these criticisms have an element of truth in Cleaver's essay; his valorization of Black lumpen revolutionary violence would contribute to his "defection" from the party after Newton's release from jail.[23]

However, the "industrial reserve army" is one of the ways he defines the lumpen in "Ideology." He writes that this Black labor reserve was made up of those "who have never worked and never will; who can't find a job; who are unskilled and unfit," as well as those "who have been displaced by machines, automation, and cybernation, and were never 'retained or invested with new skills,'" or are on welfare. Racial difference is constitutive of how the United States defines the normative, homogeneous categorization of the working class, the lumpen, and the industrial reserve army. This is why there is "confusion" about not only the political categories but also the Panthers' position on them. In Cleaver, this confusion is productive; however, it is also marked by slippage between categories; at times the categories of "working class" and "lumpenproletariat" lose their racial difference, becoming the "right wing" and the "left wing" of the "proletariat" in both the mother country and the Black colony. In contrast to those dispossessed by automation and cybernation, "the Working Class of our time has become a new industrial elite. . . . Every job on the market in the American Economy today demands as high a complexity of skills as did the jobs in the elite trade and craft guilds of Marx's time. In a highly mechanized economy, it cannot be said that the fantastically high productivity is the product solely of the Working Class. Machines and computers are not members of the Working Class, although some spokesmen for the Working Class, particularly some Marxist–Leninists, seem to think like machines and computers." Here technology moves between a formation that works in tandem with highly skilled workers, the combined productivity of the two making a nonracialized working class an "elite" social formation. This is the opposite of how the League describes the automated factory; it also does not quite capture the sense of class division within factory work as described by James Boggs. Yet Cleaver, who also critiques the mechanistic thinking of then-contemporary American Marxists, shares with Boggs and the League the sense that machines and computers are fixed capital that mobilize all of science, nature, and social difference against the proletariat, further differentiating it into those elites who have the skills to keep up with technology and those who are rendered lumpen.[24] This raises an important political question, both for thinking about the Black Panthers' developing theories of political struggle in the cybercultural era and for thinking about how their work applies to today's planetary capitalism and its proliferation of antagonisms: is the category of lumpen an economic or a political category, and does it have a racial specificity?

Throughout, Cleaver wavers on this, focusing on Black dispossession while also arguing that the working class and the lumpen in both the mother country and Black colony are part of the "Proletariat, a category which theoretically cuts across national boundaries but which in practice leaves something to be desired." Although critics might claim that Cleaver fails to offer a theoretically nuanced consideration of the "confusion" that he so effectively frames at the beginning of his essay, it instead parallels what Étienne Balibar calls Marx's "omnipresent uncertainty" about the proletariat across his work.[25] According to Balibar, the proletariat is a category that "short-circuited" politics and economics. For Marx, it was characterized not as a "class" but instead a "mass," in the sense that the proletariat is not synonymous with a working class produced by capitalist relations of production or a class defined as that which has a class interest, like the bourgeoisie, but instead as the dispossessed masses. Out of capitalism's disorganization of social life, and at the beginning of the revolutionary process, the proletariat will coalesce into a "positive universality" that will lead to the dissolution of all classes.[26]

In "On the Ideology of the Black Panther Party," Cleaver attempts to define the Black lumpenproletariat as a revolutionary mass, especially in relation to revolutionary claims by White mother country radicals. In chapter 4 I have already discussed how Cleaver analyzes the social imaginary that technologically separates White and Black, mind and body, and that was put under pressure by both technological change and changes in popular culture. Although these are very different essays, in "Ideology," race is a social imaginary made productive by capitalism to prevent class formation—a condition that is only partly exacerbated by automation and cybernation technologies. The Black lumpenproletariat is a product of both this long history and its then-contemporary specificity. Black and lumpen thus sometimes appear to be synonyms and lumpen a racial category, and at other times, race and class are independent categories. However, as a product of capitalism's continued production and exploitation of racial difference, Cleaver's Black lumpenproletariat necessarily occupies an unstable place between being an economic, political, and racial formation.

What was ultimately at stake for the Black Panthers was defining the protagonist of its anticapitalist political project at a moment in the technological reorganization of capitalism, which Cleaver believed was the Black lumpenproletariat because they were the dispossessed of the dispossessed. While many have taken the lumpen to have always been *the* revolutionary subject for the Panthers, Angela Davis openly questioned that status in a two-part conversation with Elaine Brown, "Welcome Home, Angela Davis," that was published in *The Black Panther: Intercommunal News Service* on March 4 and 11, 1972. In the first part, Brown contends that Black people, who are already seen as "criminals" instead of "workers" in a carceral "slave labor supersystem" in which they are used for surplus-value extraction, will only further be thrown into to the prison system as technological change increasingly displaces them into unemployment and the informal, potentially illegal, economy. In response, Davis says, "That's why we really have to revaluate what has traditionally been called the lumpen proletariat. There has been the tendency, as we know, to romanticize the lumpen proletariat," either as a revolutionary class or as criminals and outlaws.[27] She continues that "we probably have to redefine our terms" in order to better theorize "the economic function of racism" and to further the revolutionary cause. In this section of the discussion, Davis departs from the Panthers' valorization of the lumpenproletariat in favor of the Black worker, who, she contends, is largely isolated from political activism in a racially unequal working class. She further argues that as workers, they are separated from the Black community because the workplace-factory is posited as a realm separate from the rest of social life. This is a condition that she believed at the time was beginning to change because Black workers were increasingly at the forefront of union struggles—something that was newly visible but also continuous with a long history of Black radical labor activism.

I draw attention to this conversation for two reasons. First, although Davis departs from Cleaver's valorization of the Black lumpenproletariat as the central figure in the Black liberation struggle, Davis and Brown's conversation shares with Cleaver a methodological orientation that rethought normative Marxist concepts and the cybercultural era's political horizons of the future. In these pages, occurring as it does in the immediate aftermath of Davis's release and in relation to broader anti-incarceration struggles, Davis and Brown's reevaluation of the status of the lumpen and the worker in technocapitalism is connected to incarceration. By making this connection, they seek to traverse realms that are necessarily separated by capitalism; in the process, they draw attention to the status of the worker as well as to the ways that unfreedom and unfree labor, especially in the form of prison labor, conditions freedom and so-called free labor.

Brown and Davis are here implicitly, and to some degree explicitly, in conversation with George Jackson and his theorization of revolution, the prison, and neoslavery. The lumpenproletariat is undoubtedly a central political figure in Jackson's thought. In *Blood in My Eye,* the Black prison population is made up almost entirely of lumpenproletarians, and their criminality, alongside their tenuous relationship to the formal economy, necessarily makes them revolutionaries. As he wrote, "Revolution is illegal. . . . It is clear that the revolutionary is a lawless man. The outlaw and the lumpen will make the revolution. The people, the workers, will adopt it. This must be the new order of things, after the fact of the modern industrial fascist state."[28] There is much that can be said about lawlessness and revolution in Jackson. For example, lawlessness is a revolutionary advantage because the fascist "city-state," as he formulates it in *Blood in My Eye,* is a "cybernetic" one subject to the "science of control," which, as he sees it, produces a weakened and bureaucratic bourgeoisie unable to adapt to guerrilla struggle.[29] I want to focus here on his sense of fascism as a particular type of capitalism whose key political concept is not the lumpen but instead neoslavery, or sometimes just slavery, which emphasizes the way capitalism always requires forms of work.

Jackson's history of U.S. racial capitalism can be described in brief: "The Civil War destroyed the landed aristocracy. The dictatorship of the agrarian class was displaced by the dictatorship of the manufacturing-capitalist class. The neoslaver destroyed the uneconomic plantation, and built upon its ruins a factory and a thousand subsidiaries to serve the factory setup. Since we had no skills, outside of the farming techniques that had proved uneconomic, the subsidiary service trades and menial occupations fell to us. It is still so today. We are a subsidiary subculture, a depressed area within the parent monstrosity."[30] Contemporary dependent neoslavery traverses the "factory complex" and the prison because they are both spaces where life is regimented through forms of technological control like time management.[31] At the heart of this neoslavery is undoubtedly the prison-industrial complex, a place of forced labor for inadequate wages that was transitioning during Jackson's imprisonment from a majority White to a majority Black institution. However, in Jackson, the racialized prison plantation also becomes the organizing principle of capitalist life in general. As the above passage shows in brief, like C. L. R. James and the LRBW, Jackson believed that there was structural continuity between the plantation and the factory. Jackson argued that neoslavery was a generalized organization of lifetimes, down to the minutiae of everyday working life, where for a wage laborer, "the control of your eight or ten hours on the job is determined by others," and transport time, eating, dressing, and every aspect of daily life become a "factor in efficiency" in the working day.[32] Alternatively to the League, Lukács, and the Italian *operaismo* and autonomists, who argued for a factory model of society, Jackson, in his theorization of the prison as a countersite of intense study and politicization, suggested a neoslavery organization of society based on the plantation. In doing so, he drew attention to the ways that even forms of pleasure and the economic advantages accrued through the wages of whiteness are embedded in a broader world of unfreedom that requires the "overthrow of all existing property relations" in order to destroy the property regime of neoslavery.[33]

Undoubtedly, the lumpen are for Jackson important revolutionary protagonists. However, his theorization of neoslavery calls into question the place of lumpenization in racial capitalist development. Against arguments that automation was potentially producing mass unemployment, Jackson suggests that capitalism is a generalized technological organization of life, developed out of the logics of the plantation, whereby all of one's life is put to work for capitalism, whether that be Black prison labor, White unionized factory workers, corporate bureaucrats, or the free time outside of labor time. Jackson at points retains the category of the lumpen, but he also suggests that lumpenization never fully happens because of the extent to which life is always put to work in one form or another by, and for, capitalism.

This position closely relates to parts of Brown and Davis's discussion, which itself develops ideas on labor, race, and class that Davis began in her 1971 essay "Reflections on the Black Woman's Role in the Community of Slaves" and that she would most fully develop in *Women, Race, and Class,* especially the chapter "The Approaching Obsolescence of Housework: A Working-Class Perspective." "Welcome Home, Angela Davis" largely condenses the major argument of both of those pieces that, contrary to claims at the time by White feminists, women's oppression is not transhistorical. Instead, its specificity is constituted not just by gender but also by a racialized and classed organization of the relations of production in a broad sense, which requires multiple forms of work by Black women and in turn has a number of political consequences. In the context of the Black Panthers' cybercultural politics, this raises the question of whether Black women can be lumpenproletariats because of their relationship to work. Davis suggests not.

In the "Obsolescence of Housework" she famously writes:

> One of the most closely guarded secrets of advanced capitalist societies involves the possibility—the real possibility—of radically transforming the nature of housework. A substantial portion of the housewife's domestic tasks can actually be incorporated into the industrial economy. In other words, housework need no longer be considered necessarily and unalterably private in character. Teams of trained and well-paid workers, moving from dwelling to dwelling, engineering technologically advanced cleaning machinery, could swiftly and efficiently accomplish what the present-day housewife does so arduously and primitively.[34]

She attributes this "closely guarded secret" to the structural requirements of capitalism to create forms of separation, such as the domestic sphere in which labor power is reproduced through the largely unwaged (or poorly paid) and gendered as feminine labor, and the economic sphere where labor power, gendered as masculine, is waged. For Davis, technology is machinery, but it is also a racialized and gendered form of organization. In *Women, Race, and Class,* she charts the ways that early industrial capitalism created the reproductive–productive split by moving production outside of the home and into the factory/city. This itself is part of a general technological ideology of capitalism in which the human political-economic world is produced as separate from nature while dialectically women are externalized as nature. As she argues in "Women and Capitalism: Dialectics of Oppression," if men, through industrial technology, become masters of nature, then women are projected as "embodiments of nature's unrelenting powers."[35] This is, of course, riddled with contradictions. In terms of the technologies of wage labor, the "mechanized" assembly line perhaps most fully represents the capitalist promise to create a "universal equivalence" of abstract labor subject to the machine.[36] Although, as the League and others demonstrate, fixed capital as mechanization is mobilized against the worker, deskilling and (ideally) depoliticizing them, it also potentially offers, or at least claims to offer, a postlabor and postscarcity world. Davis argues here that gendered difference, like race, is mobilized throughout production and social reproduction to undermine this "rationality" and its immanent potentials.

The central thread that connects the ideas that Davis and Brown discussed in *The Black Panther: Intercommunal News Service* to the technologies of housework in *Women, Race, and Class* is the valorization of living labor and the active struggle of Black women for equality through the recognition of their lives as workers. To do so is historically fraught, given the ways that work, gender, and race are conditioned by a history of slavery in which on the plantation, "women, no less than men, were viewed as profitable labor-units[;] they might as well have been genderless as far as the slaveholders were concerned"—except when they were subjected to gendered forms of violence.[37] The inequalities of the afterlife of slavery required that, unlike middle-class White women, Black women continued to be workers, a condition essential to the reproduction of the Black community. With this history in mind, in the early pages of *Women, Race, and Class,* Davis places special emphasis on Marx's concept of living labor as an active creative force; it is their existence as living labor that gives Black women "a confidence in their ability to struggle for themselves, their families and their people."[38]

Saidiya Hartman has cautioned against placing even more weight on the shoulders of Black women as the bearers of freedom.[39] Yet for Angela Davis and Elaine Brown, like for activists in the NWRO, Black women as workers was a political concept that brought to the fore the extent to which acknowledged and unacknowledged forms of racialized and gendered work are required to make capitalism function practically and ideologically, and it foregrounded the extent to which Black women were central to the struggle for freedom. To return to the starting point of this discussion, by questioning the conceptual role of the lumpen, Davis foregrounds not just the importance of Black workers but also the ways that the lumpen is a gendered category that excludes Black and other oppressed women because of the centrality of work in all of its waged and unwaged forms. Following Davis, Brown, and Jackson, as well as Cleaver, we can further say that the concept of the lumpen is productively confusing, especially at a cybercultural moment of technological change, because it foregrounds the extent to which capitalism continues to demand that work structure life and not be freed of it.

The composition of work and class, along with gender and race, has so far been at the heart of the debate over the lumpen. Huey

Newton's theory of intercommunalism, however, shifted it to a different terrain. In the tradition of Hegel, Lenin, and C. L. R. James, Huey P. Newton was a dialectical theorist of the "leap."[40] In *In Search of Common Ground,* Newton argued that "class conflict develops by the same principles that govern all other phenomena in the material world" and that although one cannot predict how and when capitalist social relations will be transformed, it is sure, as he says, "that if we increase the intensity of the struggle, we will reach a point where the equilibrium of forces will change and there will be a qualitative leap into a new situation with a new social equilibrium."[41] As a theorist of the leap, Newton argued that the Black Panthers' ideology had been out of sync with actual political developments. Previously the Panthers saw themselves as either revolutionary nationalists or internationalists; however, because the world was no longer constituted by sovereign nations but instead "empire," the anticapitalist struggle for equality could no longer be led by the lumpen and wage its battles at the level of the (nonexistent) nation-state. Instead, it must be a "community" undertaking.

As this suggests, Newton's dialectical materialism of the leap understood the world to be constituted by continual change driven by contradictions. Put succinctly, the world is revolutionary. As he told the audience at the Revolutionary Intercommunal Day of Solidarity on March 5, 1971, "We must remember that Revolution is a process. It's not a conclusion; because once we conclude, then we become counter-revolutionary. . . . We always have welcomed all forms of contradictions. Because without the contradiction there is no transformation. . . . But we must not stop the Revolution. . . . So we're ideologically revolutionists. We're in a constant state of change."[42] This included Newton's approach to the Black lumpenproletariat. Although popularly valorized, starting in late 1970, Newton argued that they were a minority, and with that reconceptualization, Newton abandoned any immediate hope for armed revolution. Instead, as he saw it, the lumpenproletariat would be a central figure in a capitalism to come:

> While the lumpenproletarians are the minority and the proletarians are the majority, technology is developing at such a rapid rate that automation will progress to cybernation, and cybernation probably to technocracy. . . . If the ruling circle

> remains in power it seems to me that capitalists will continue to develop their technological machinery because they are not interested in the people. . . . If revolution does not occur almost immediately . . . the proletarian working class will definitely be on the decline because they will be unemployables and therefore swell the ranks of the lumpens, who are the present unemployables. Every worker is in jeopardy because of the ruling circle, which is why we say that the lumpenproletarians have the potential for revolution.[43]

The lumpen, in Newton's analysis of the cybercultural conjuncture, is a "potential" revolutionary subject against technocracy. Although not fully defined here, technocracy is a capitalistically administered planetary social relation based on imperial expropriation of the material, intellectual, and cultural resources of the intercommunal world. Technocracy is the logic of what Newton called "empire," or "Reactionary Intercommunalism," which John Narayan and Delio Vasquez argue anticipates Michael Hardt and Antonio Negri's *Empire* in most of its political contours.[44]

Intercommunalism was the fullest expression of Newton's theorization of the logic of racial capitalism. It is also arguably the Black Power cybercultural concept that most fully captures the transition from the Fordist–Keynesian organization of production and reproduction into a neoliberal one while foregrounding the mutually constitutive role of racialized social difference, technological change, and imperialism to that process. Newton argued that the "ruling circle" of the United States, or "North America," had undermined the sovereignty of nation-states and fully reorganized capitalism on a "world-wide level" through a combination of the military that facilitated political control, the mass media and transportation technologies that had collapsed space and time, and the technologies of resource extraction. Extraction and appropriation for Newton were not limited to raw materials or natural resources but also included the cultural and intellectual resources of colonized and (formerly) enslaved peoples. In dialogue with theories of underdevelopment and accumulation through dispossession, technology was for Newton a political, imperial formation whereby appropriation created a "reservoir of information" for capitalism, which in turn created the "technological machine" of political and institutional forms through which empire itself is created.[45] Newton called this way of seeing "technology" and responding to it "the technology question."[46]

Intercommunalism and empire as technology questions raised important issues, especially for thinking about the logics of the cybercultural revolution at the end of the Black Power period. First, because "you cannot change a part of the whole without changing the whole," empire as a new organization of planetary capitalism radically alters forms of sovereignty, meaning that the nation-state was no longer the dominant political form.[47] Newton theorized, at the end of the war in Vietnam, that the United States remained a "sovereign stronghold." However, he meant this strictly in the sense that from there, a tiny number of corporate capitalists were able to mobilize the U.S. military in order to secure the planet for the "long arm" of empire.[48] Newton's conception of empire anticipates in many respects Michael Hardt and Antony Negri's definition of empire as a new global form in which capitalism's increasing ability to move across nation-state borders means that there is no longer an imperial center but instead a new world without boundaries and in which hybrid forms of sovereignty seek to manage not territory but human nature.[49] However, for Newton, the United States remained, at least for the moment, the mechanism for enforcing the new planetary form of capitalism. In this respect, his sense of nation-state sovereignty in transition was closer to Wendy Brown's argument that today, the nation-state's role has not disappeared but is instead undercut by a "neoliberal rationality" that the state enforces to undermine democratic politics.[50]

Second, consistent with his early arguments in favor of lumpen, Black nationalist, and internationalist politics, Newton was primarily concerned with identifying a political strategy for creating economies of equality. If the nation-state was no longer the dominant political form, then nationalist and internationalist politics no longer made sense. Instead, he argued for a "revolutionary Intercommunalism," with the primary unit of political organization being the "community."[51] Newton defined the nation, "as opposed to a community," as a group of human beings that "have in common their own land or territory, economic system, culture (or way of day-today living), and language, etc."[52] Newton moved away from his previously held nationalist positions because he argued Black people in the United States did not have a nation in this sense, or indeed a "superstructure of our own. The superstructure we have is the superstructure of Wall Street."[53] Instead, he defined a community "as a small unit with a comprehensive collection of institutions that exists to serve a small group of people . . . [and] who want to determine their own destinies," and that could potentially have the same function as the state in the institutional sense.[54]

If the lumpenproletariat in the Black Panthers' thought is best described by a great degree of "confusion," *community* is in general also a fraught term with multiple and conflicting meanings, with its power coming from its rhetorical flexibility. This is largely true in the Panthers' work, where it appears too often to be contextualized. As an explicit political concept in dialectical relation to imperialism and colonialism, Eldridge Cleaver argued in 1968 in *The Black Panther: Black Community News Service* that Black people were subject to "community imperialism" in which the Black community/colony was "organized" into poverty by White "mafia"-style capitalism that controlled the community from the outside. "Community liberation" was thus a necessary step in the struggle for "National Liberation."[55] Newton's theory of intercommunalism argued, alternatively, that community emerges as a political category not so much as an alternative to nation-state dominance but as an effect of its dissolution.

Miranda Joseph argues that "community" is not a natural coalescence defined by internal unity in opposition to an outside; instead, it is produced through the power dynamics of inclusion and exclusion. She further contends that although communities are usually posited as capitalism's other, they in fact "supplement" capitalism and are complicit with it by doing the work of care and, tautologically, community, which capitalism attempts to undo, and they do so often through consumerist practices of self-definition.[56] Newton's, and Cleaver's, sense of community politics shares something of this in that it is posited in both formulations as an opposite to capitalism. In Newton's case, it is thus required to do the care work, through survival programs, that a deterritorializing capitalism was increasingly abandoning. Yet if capitalism universalizes, in the bad sense, the commodity form and the consumer subject, and it displaces the role of the welfare state onto private groups like the Panthers, John Narayan argues that unlike Hardt and Negri's multitude, or the everyday sense of community as natural unity, Newton's politics of community directly confronts the existence of difference and the ways that it is produced. Communities are locations of difference interlinked through political struggle—not to create a utopia removed from contradiction but instead to create cooperative forms of ownership through which infrastructures can be built to ultimately overcome inequality and create a new, positive, and noncapitalist universality.[57]

I trace the confusing nonlinear trajectory of the Black Panthers' theorization of the political protagonists of the cybercultural revolution, but not to claim that any of these protagonists offers the correct answer. Instead, turning to their struggle over the lumpenproletariat demonstrates the nuance of their collective theorization and the political dynamics through which they, and more specifically Newton, arrived at the concept of community that was the protagonist through which they tried to organize alternative forms of social reproduction in the 1970s, one of which was to be people's community control of modern technology.

Survival and Reconstitution

I began *On the Eve of the Cybercultural Revolution* by arguing, through an analysis of Martin Luther King Jr. and *The Triple Revolution*, that the cybercultural era was a crisis in the production and reproduction of capitalism at a moment in which automation and cybernation technologies appeared to be reorganizing race, class, and labor in ways continuous with the afterlife of slavery, even if it was otherwise in many respects new. King responded to this by attempting to organize a mass movement that would expand and deepen the welfare state against its historical tendency toward racialized and gendered forms of exclusion. Initially, the Black Panthers shared this orientation. This is evidenced, for example, in the Ten Point Program, which demanded that the federal government provide full employment to all Black *men* and/or a guaranteed income—demands that, for a time at least, were mainstream across the political spectrum. In his 1969 conversation with Lee Lockwood while exiled in Algiers, Eldridge Cleaver claimed that the Black Panthers' political project was to create a "Yankee-Doodle-Dandy version of socialism."[58] This U.S.-specific socialism would require cooperative control of the means of production so that all could benefit from the abundance produced by human labor and technology. Cleaver also argued, however, that this socialism must take into account that the United States was a uniquely affluent country (including among the poorest of the poor) compared to socialist states, and that it was a postcolonial country that had a long tradition, or at least ideology, of individual rights and freedoms that should be further valorized, even with collective ownership and economic equality. This, for Cleaver, could only be achieved through interracial armed revolution against the U.S. government and its policing technologies, in conjunction with a planetary anticolonial and socialist struggle. As this indicates, while technocapitalism was understood to be a (neo)colonial organization of life, it was through the nation-state that equality through technology was to be achieved.

Intercommunalism marked an end to the Panthers' immediate focus on armed revolution, and at least theoretically the nation-state, in favor of community politics. To return to the chapter's opening question, what did this mean for people's community control of modern technology? This is an especially important question to end the book with because the early twenty-first century is a moment marked first by the return of the automation discourse and its politics of inequality, and second by the apparently limited potential of the nation-state to secure a more just distribution of society's surplus. A potential politics of the nation-state should not be abandoned, especially if it can provide for political-economic equality and well-being, but it is necessary to ask whether, in a world in which capitalist automation technologies are deployed as a planetary formation and thus the social relations of technocapitalism are planetary, whether the nation-state is the level at which this equality, whether through UBI or other guarantees, can be, or is best, secured. The Panthers' intercommunalism offers one approach to this question.

Elaine Brown has shown that the development of the theory of intercommunalism, the formalization of the survival programs (especially in Oakland), and the Revolutionary People's Constitutional Convention (RPCC) of 1970 were closely linked.[59] Of these, the survival programs are undoubtedly the best known and most revered of the Panthers' accomplishments. They included the Intercommunal Youth Institute; Seniors Against a Fearful Environment, better known as SAFE; the Free Food and School Children's Breakfast Program; the People's Free Shoe and Free Clothing Programs; and the People's Free Medical Research Health Clinics and the Sickle-Cell Anemia Research Foundation. Alondra Nelson has called programs like the latter forms of "social health" that "scaled from the individual, corporeal body to the body politic in such a way that therapeutic matters were inextricably articulated to social justice ones."[60] Robyn Spencer argues that, especially through the leadership of Panther women, with the survival programs, "they experimented with truly collective organizational structures for health, child care, and education."[61] During a period of extensive and detrimental reorganization of the welfare state and a popular emphasis on Black capitalist programs, the Panthers' "engagement with urban renewal, alternative education, and community control reflects a continuation of their quest for Black Power."[62]

Despite this, Newton famously described them as "satisfy[ing] the deep needs of the community but they are not solutions to our problems. That is why we call them survival programs, meaning survival pending revolution."[63] Although the most prominent member of the Panthers, Newton did not represent the entire group, and his position obscures the importance of, for example, Elaine Brown and Ericka Huggins in developing the party's community politics. Yet in order to understand what is at stake in people's community control of modern technology, I want to stay here with Newton's conception of the survival programs as part of a long-term process challenging a complex technological reactionary intercommunalism. In "The Technology Question," Newton noted that "we, the people, do not have a worldview."[64] This echoed what he claimed was the true intention of the survival programs. They were "not answers or solutions, but they will help us to organize the community around a true analysis and understanding of their situation. When consciousness and understanding is raised to a high level then the community will seize the time and deliver themselves from the boot of their oppressors."[65] In order for the community to fully control technology and to create new forms of social reproduction, they thus required a "high level" of knowledge that located technology, and their constitutive relation to it, as an extractive reservoir of information for planetary capitalism. Intercommunalism was not only a shift from the lumpen to the community required by a new analysis of revolutionary cybercultural capitalism reorganizing itself as empire, but also an ethical-political project that argued that a new social formation, a "community" initially made by the technologies and logics of empire, could (re)constitute itself out of this reorganization of capitalism.

Much like Friedrich Engels, Newton wrote in his own "Dialectics of Nature" that intercommunalism was premised on "the basic concept of the unity of nature underlying and transcending all arbitrary national and geographic divisions" that was "confirmed" by science.[66] Writing at a moment, like today, in which mainstream environmentalism often posited a split between society and nature, he argued in the tradition of Engels, as well as the Frankfurt School and Angela Davis, that this split was the result of a capitalist project to reconstitute humanity's relation to the world. While science could demonstrate a unity of nature that traverses false species and political boundaries, technoscience, as a planetary formation, "slaves away in the service of reactionary intercommunalism"[67] and aids the capitalist-imperialist "madmen" who seek to "destroy nature" in accomplishing two goals: (1) to divorce humans from a "rational relationship with the nature world"[68] in order to (2) remake humanity into consumers of the products of extractivist capitalism. Newton ultimately urges his readers to reorient their thinking in a dialectical way: to see crises like ocean pollution, shale oil and coal extraction, and the use of napalm as both environmental and political crises.

Written while a PhD student at U.C. Santa Cruz, "Dialectics of Nature" was one in a series of works across media where Newton confronted a major political problem. In a complex world of mass consumerism, information reservoirs, and metabolic rift, a person's location in the system was "hard to pinpoint." This was especially true for Black and working-class people ("hard hats") in the United States who needed to earn a living and who wished to benefit from increased wages and access to consumer goods, even though they were produced through the domination of nature, harm to Third World peoples, and increasing racialized inequality.[69] Through writing, organizing, and the survival programs, Newton responded to this by politicizing technoscience, reorienting it from a strictly specialist undertaking into a popular "activist" approach to knowledge that dialectically "traces the poisonous spoor" of racial inequality and ecological damage to the ways capitalism organizes human and extrahuman nature as a reservoir for extraction.[70]

Perhaps the most interesting attempt to instill this form of consciousness was instituted in the curriculum of the Intercommunal Youth Institute, which was directed by Ericka Huggins.[71] In the fall 1974 issue of Stewart Brand's *CoEvolution Quarterly* that the Black Panther Party guest edited, the Panthers described how they organized children's education. Under the heading "Matter and Energy," they grouped the study of "Science," "People's Art," and "Political Education." They wrote that the goal of science classes was to understand "the balance between man and nature as well as other scientific knowledge necessary to exist in a highly technological, scientific world," and the goal of political education was to "be critical of the situation the world is in, to foster an investigative attitude, to provide a framework for the comparison of different peoples and their politics."[72] If interlinking these in a single curriculum might evoke Jorge Luis Borges's "certain Chinese Encyclopedia" that Foucault claimed "shattered" normative models for organizing knowledge, they are in fact a way of politicizing the organization of STEM that lays a foundation for "People's Community Control of Technology."[73] In order to control technology, the community must see it not as simply machines to be manipulated but as a form through which reactionary intercommunalism has produced multiple rifts in the service of organizing life as reservoir. Education and the survival programs as consciousness programs attempts to respond to this "dilemma" through "control of that vehicle to which we have all contributed through the exploitation of our lives and labor."[74]

As the last line suggests, consciousness means nothing without the ability to control the technologies of extraction and dispossession. The Ten Point Program already argued that robust political change was necessary to ensure rights already guaranteed by the existing U.S. Constitution.[75] However, with the RPCC in Philadelphia in September 1970, they went a step further, arguing that formal reconstitution was necessary in order to address the material, economic, and political changes that had taken place since 1789 and to give oppressed peoples greater control over their lives. Although organized by the Panthers, the event involved participants from across the radical spectrum, including representatives from the American Indian Movement, the Young Lords, the Gay Liberation Front, and various feminist groups. While the numbers vary, somewhere between five thousand (according to *The New York*

Times) and fifteen thousand people (according to the organizers) attended.[76]

Above all, as participant George Katsiaficas argues, this was an "ordinary people's" convention, not simply a showcase for revolutionary leaders.[77] Although there were speeches by Michael Tabor, Newton, and others, it was the convention's working groups that produced a document of principles that was ultimately supposed to be shaped into the final constitution at a second session in Washington, D.C., in November. These included the "Workshop on Internationalism and Relations with Liberation Struggles Around the World," the "Self Determination for Street People" working group, the "Workshop on the Self Determination of Women," the "Statement of Demands from the Male Representatives of National Gay Liberation," and the workshop on "The Family and Rights of Children" (which included no children, as the session notes state). They produced a wide-ranging set of principles for the reconstitution of American society, including making illegal the mass media's "exploit[ation] of women's bodies in order to sell or promote products," guaranteed nondiscriminative employment in workplaces controlled "collectively by the working people," guaranteed prepaid maternity leave, socialization of housework and a guaranteed income, abolition of the nuclear family, "the right to be gay anytime, anyplace," the modification of language "so that no gender takes priority," technological liberation from "the world of drudgery," "free de-centralized medical care," the legalization of psychedelics (although no "revolutionary action should be attempted while under the influence" of them), and the formation of "revolutionary people's community councils" to implement the new constitution. In addition was the first demand in the document as it has been preserved by Katsiaficas: the right of the United States to "nationhood" was denounced because it had become "an international federation of bandits."[78]

According to Katsiaficas, the final session never fully happened as planned because of Newton's lack of interest and because of the split between the Newton and Cleaver factions of the party.[79] Although the RPCC was perhaps always a romantic idea, despite claims by Elaine Brown, David Hilliard, and Katsiaficas, charting Newton's own statements demonstrates that he placed a great deal of importance, at least theoretically, on reconstitution and its relationship to his developing ideas about technology and intercommunalism. Newton's first statement on the convention came in a *Black Panther: Black Community News Service* article on June 13, 1970, entitled "Towards a New Constitution."[80] There he argued that a constitutional convention was "extremely important" because it would clarify and popularize to the world the Panthers' struggle against "bureaucratic capitalism and American Imperialism" and it would formalize the party's demand for a constitution that reflected the "pluralistic" nature of U.S. society. Beyond this ethical imperative, the need for a new constitution was also necessitated by the ongoing material reconstitution of capitalism. If the 1789 formal Constitution was based on a mode of "land and soil" where "democratic capitalism" was available to White property-owning men, by 1970, U.S. capitalism had become, according to Newton, "overdeveloped." This overdevelopment meant that in a new planetary capitalism, the United States was investing its surplus capital in other parts of the world, not domestic Black and oppressed communities. It also meant that there was no more room "within" capitalism for Black autonomous development—if in fact they had surplus capital to invest. In "Towards a New Constitution," Newton has yet to fully make the intercommunal leap, arguing instead that proportional representation for racial and ethnic groups in a "socialist framework" was the only way to ensure human rights for Black people in the United States.

By late 1970, Newton's early theorization of intercommunalism was closely tied to the constitutional convention process. At Boston College on November 18, 1970, a speech reproduced in *The Black Panther* on January 23, 1971 (right before it became the *Intercommunal News Service*), Newton began his discussion of intercommunalism with the RPCC, which he argued was necessary because there, the Panthers would introduce their "total survival program," thereby suggesting that the convention was only one part of the revolutionary intercommunal process. I have already analyzed his main argument from this speech about the transformation of planetary capitalism from a collection of nation-states into a U.S.-led empire that not only dispossess intercommunal peoples but also mobilizes technology to collapse space and time while dialectically transforming the world into a collection of communities that appear fragmented despite their integration through exploitation.

In this speech and the one that followed at the RPCC, Newton attempts to construct this picture of totality from the perspective of Black and intercommunal peoples in order to explain why reconstitution was necessary.[81] Newton makes this explicit in his "Resolutions and Declarations" at the anticlimactic final November session of the convention: "This Convention of Revolutionary Peoples from oppressed communities throughout the world is convened in recognition of the fact that changing social conditions throughout the world require new analyses and approaches in order that our consciousness might be raised to the point at which we can effectively end the oppression of people by people."[82] In what would have been the first place, outside of Boston College, that the public would encounter intercommunalism, he argues here that formal reconstitution is explicitly required because of the material reconstitution of planetary capitalism, which has largely been misidentified, according to Newton, by revolutionary activists and the Black community, too focused as they are on the nation-state and state capitalism. Katsiaficas contends that during the constitutional process, Newton, in part because of government repression, undermined the "multicultural" nature of the RPCC by focusing on the survival programs of the Black community.[83] At the same time, Katsiaficas argues that Newton's theory of intercommunalism marked a shift away from the values represented in the constitutional framework. To some extent this may be true, although the shared critique of the U.S.'s national status demonstrates otherwise. Intercommunalism, as it is presented in relation to the constitutional process in his November speech, suggests a more profound rethinking of the politics of constitution. On the one hand, the original formal U.S. Constitution was based on slave labor and labor-intense agricultural technologies, and it was written at the emergence of the nation-state as a legal and political form in which White landowners also controlled geographic territory. Newton argued that the Keynesian state management of capitalism, which in the United States was an extension of this original framework, could no longer provide for the reproduction of nation-state capitalism because it had been exceeded by empire. This required a domestic reorganization of the constitutional framework of American life. Newton also makes a broader and more provocative statement: "We are here gathered for the solemn purpose of formulating a *new constitution for a new world*."[84]

The conventioneers must be conscious, he said, of how this world is constituted politically and economically. They must "[make] use of all the technology and knowledge we have accumulated" and produce a constitution that is properly intercommunalist—that is to say, a constitution not based on national sovereignty defined by territorial boundaries but one that recognizes the rights of every community on the planet to "define, determine and control [their own] institutions" in a framework that provides for fully democratic participation by those communities and that thus necessitates more-than-capitalist legal and political structures.[85]

Throughout this chapter, I have sought to show how the Black Panthers were engaged in a capacious theoretical undertaking to understand the forces driving the cybercultural revolution—but also to understand what political practices were most appropriate to the conjuncture. This in part meant continually grappling with who the most appropriate protagonists were to lead the anticapitalist revolution. Was it the lumpenproletariat? The worker? The community? Continually rethinking the revolutionary subject was intertwined with rethinking the geographies of capitalism (and anticapitalism), but also its techniques and its political and institutional forms. As I have argued throughout, central to all of this was theorizing the ways that technology was an organizational form tied to policing and control, but especially to extracting value from racialized and oppressed peoples and nature more broadly. In light of this, the RPCC thus suggests that what is at stake in the people's community control of technology was not simply localized in Oakland and other Black communities. Instead, it was tied to the ways that capitalist technologies were reorganizing extant political structures, and in response required an expansive political process to redefine the modalities of power and social reproduction through which oppressed peoples and communities participated in the world.

Huey P. Newton's "Resolutions and Declarations" speech ends in a surprising place for a text about writing a new constitution. He claimed that technology and knowledge will only properly be put to use in the "true interests of the people" when a constitution of revolutionary people has fully "developed a system" in which "the word 'work' will be re-defined as meaningful play."[86] In these words we might hear echoes of the Panthers' *détourning* of the Declaration of Independence's call for life, liberty, and the pursuit of happiness. If this was Newton's intention, it was reworked for the specificities of the cybercultural era and its overriding concern with how automation and cybernation were reorganizing race, labor, and class in ways that were creating new forms of inequality (even if historically continuous with the afterlife of slavery) while also potentially creating a new postlabor, postscarcity world. In this respect, it is more than appropriate to end with Newton's belief that a fully realized constitution would not only provide for the full participation of all communities in the world-making process but would also utilize technology "expropriated from the expropriators" to create new forms of play and thus freedom.

On the Eve of the Cybercultural Revolution began with Martin Luther King Jr.'s engagement with *The Triple Revolution* and his call for an expansive welfare state, moved through the heterogeneous political-economic perspectives that constituted Black Power cyberculture, and concluded with the RPCC's struggle with what form a new constitution, attuned to the contradictions of the cybercultural revolution, should take. These contradictions are not the contradictions of the present. Yet taken as a whole, I have tried to show that Black Power cyberculture offers a particularly rich theoretical perspective on the cybercultural organization of life as well as on the political possibilities that could have emerged from that moment, and that may still offer counterpractices for the creation of new values alternative to racial capitalist automated inequality today. In this respect, Black Power cyberculture's horizon of the future remains open. To reiterate the words Newton spoke at the Revolutionary Intercommunal Day of Solidarity on March 5, 1971: "We must remember that Revolution is a process. It's not a conclusion; because once we conclude, then we become counter-revolutionary. . . . We always have welcomed all forms of contradictions. Because without the contradiction there is no transformation. . . . But we must not stop the Revolution. . . . So we're ideologically revolutionists. We're in a constant state of change."[87]

Acknowledgments

In her acknowledgments for *Golden Gulag,* Ruth Wilson Gilmore writes about all of the debts one accumulates when writing a late first book. This is definitely the case for me, and her words have been on my mind. This book has taken a long road to publication; it contains things I've thought about while at Arizona State, San Francisco State, Columbia University, and the bookstores of San Francisco. Along the way, too many people to name have shaped my thinking; I wish I could thank them all. I acknowledge the profound debt that I have to the subjects of this book, the scholars I cite, and the many people whose work has influenced me but who I don't cite directly. I have tried as a white scholar, teacher, and person to learn from many different Black, radical, and countercultural traditions, as well as those people known and unknown to me who continue to create them. Despite the many ways that this could be a better book, I hope that my attempt here to honor those debts is evident.

I thank Brent Hayes Edwards, David Scott, and Maura Spiegel, as well as Kellie Jones and Jordan T. Camp, for reading this book in an early iteration and for offering their generous feedback. Thanks especially, Brent, for your continued support since I first visited Columbia. Thanks as well to Neda Atanasoski and Jonathan Flatley for their early feedback and to Neda for also reading the whole thing and offering critical and supportive insights. Thanks to readers at the Diaspora, Race, and Empire colloquium as well as *Cultural Critique, Rethinking Marxism,* and *Journal of Literary Studies* for helping to make earlier versions of some of these chapters much stronger. I also had the opportunity to present on panels at the Getty Research Institute, the American Comparative Literature Association, the Society for Literature, Science, and the Arts, the American Comparative Literature Association, the Modern Language Association, and the University of Lisbon Center for English Studies. Thank you to my fellow panelists and to audience members' questions that helped me to refine my ideas. Thanks to Jane McFadden and Kathryn Leonard for the opportunity to teach courses at ArtCenter and Occidental related to this project.

At the SLSA, I met Doug Armato of University of Minnesota Press. It is only because of that chance meeting, his encouragement to send him a proposal, and his support throughout that this book has come to be. I have always wanted to publish with Minnesota, and to see this actually happen is a special thrill. Thank you Doug. I also thank Shelby Connelly, Rachel Moeller, and Mike Stoffel. Thank you Karen Hellekson for editing and improving the text. Thank you especially Zenyse Miller for all of your guidance and help along the way.

In the introduction, I write that the motivation for this book was thinking about the relationship between James and Grace Lee Boggs's and Noah Purifoy's simultaneous, but different, concerns with value. However, in life, it began with Nijah Cunningham's suggesting that I read "The Negro and Cybernation." Over time, the Boggses have become even more important to the book, as well as to how I think about my activities more generally, and without his suggestion, I have no idea what direction this project would have taken.

Most important, this book is one moment in a long-term process that would not have been possible without many emotionally and intellectually supportive relationships. I cannot thank everyone enough, including many not listed here. However, thank you especially Ati Akbari, Valerio Amoretti, Steve Bartell, Stéphane Benoist, Campbell Birch, Marie Buck, Joe Cady, Christopher Chekuri, Nick Croggan, Andy Crow, Dan Field, Trevor Getz, Nicole Gervasio, Cristina Pérez Jiménez, Christopher Knox, Bret Maney, Megan Martenyi, and Alex Pittman. Thanks to the workers in the Occidental College NTT Faculty Union, the ArtCenter Faculty Federation, the Student Workers of Columbia, and to the other workers and unions on those campuses. Thank you John Elrick for so many conversations related, and unrelated, to this book.

I wrote this as a contingent academic worker, and I received no funding to complete it. Although I owe them beyond anything having to do with this book, any time I was able to take away from teaching to dedicate myself to it is because of three people. Thank you to my parents, Fred and Toni, for your love and generosity. Casey, we've been through so much together, it's impossible to say everything here that I'd like to. I can't thank you enough for supporting me during not only the writing of this book but through everything else. With love from the bottom of my heart, thank you. I could not have done this without you.

Notes

Introduction

1. James Boggs, "The Negro and Cybernation," in *The Evolving Society: Proceedings of the First Annual Conference on the Cybercultural Revolution—Cybernetics and Automation,* ed. Alice Mary Hilton (Institute for Cybercultural Research, 1966), 168.

2. James Boggs and Grace Lee Boggs, *Revolution and Evolution in the Twentieth Century* (Monthly Review Press, 1974), 19.

3. Throughout, capitalization of *White* signifies that it is a political category of inequality and hierarchical social difference (and, in certain places, an identification with White Power). Alternatively, capitalization of *Black* signifies that it is a term of identification, especially in relation to the Black Power movement. I also occasionally may not capitalize either when it seems appropriate, usually when signifying processes of racialization. Of course quotations reproduce the spellings used by the writers, which reflect their era.

4. Alys Eve Weinbaum, *The Afterlives of Reproductive Slavery: Biocapitalism and Black Feminism's Philosophy of History* (Duke University Press, 2019).

5. James Baldwin, *No Name in the Street* (Dial Press, 1972), 53.

6. Neda Atanasoski and Kalindi Vora, *Surrogate Humanity: Race, Robots, and the Politics of Technological Futures* (Duke University Press, 2019), 4.

7. See Grace Lee Boggs, "The City Is the Black Man's Land," in *Living for Change* (University of Minnesota Press, 2016); and the essential biography by Stephen M. Ward, *In Love and Struggle: The Revolutionary Lives of James and Grace Lee Boggs* (University of North Carolina Press, 2016).

8. Louis Chude-Sokei, "Race and Technology: A Creole History," *Technosphere Magazine,* April 15, 2017, technosphere-magazine.hkw.de.

9. Boggs, "The Negro and Cybernation," 171.

10. Boggs, 169.

11. Boggs, 171, 170.

12. James Boggs, *A Job Ain't the Answer* (N.O.A.R., February 1981).

13. James Boggs, *Pages from a Black Radical's Notebook: A James Boggs Reader,* ed. Stephan M. Ward (Wayne State University Press, 2011), 110.

14. Aaron Benanav, *Automation and the Future of Work* (Verso, 2020).

15. To take one representative example, the foreword and introduction to the 2006 edited volume *Critical Cyberculture Studies* virtually drops the term *cyberculture* in favor of *internet studies,* signaling that it is only through the

internet that the cyber can be known. David Silver and Adrienne Massanari, eds., *Critical Cyberculture Studies* (NYU Press, 2006).

16. This is true of classics in the field of cybercultural studies: see Mark Dery, *Flame Wars: The Discourse of Cyberculture* (Duke University Press, 1994). In this volume, he coins the term *Afrofuturism;* but with the exception of a discussion about the work of Black science fiction author Samuel R. Delany, the Black Power movement and especially the 1960s sense of cyberculture are absent. This absence is more surprising in books that closely examine the cybernetic cultures of the 1950s and 1960s, beginning with Fred Turner's essential text on the development of internet culture in the late twentieth century, *From Counterculture to Cyberculture: Stewart Brand, the Whole Earth Network, and the Rise of Digital Utopianism* (University of Chicago Press, 2006); N. Katherine Hayles's classic *How We Became Posthuman: Virtual Bodies in Cybernetics, Literature, and Informatics* (University of Chicago Press, 1999); and Ronald Kline's more recent history, *The Cybernetics Moment, or Why We Call Our Age the Information Age* (Johns Hopkins University Press, 2015). Much of Turner's book, for example, looks at the countercultural milieu around Brand and the *Whole Earth Catalog,* one that Hilton and the Boggses were adjacent to and that the Black Panthers associated with directly; see the special issue guest edited by the Black Panther Party, *CoEvolution Quarterly* 3 (1974). In his book, Turner presents a linear narrative of the transformation of the counterculture into cyberculture, describing the *Whole Earth Catalog* and its online instantiation, WELL, as marking the transition from a "countercultural critique of hierarchical government and its celebrations of cybernetic forms of collaborative organization" into "a world of individuals and organizations linked by networks of computers—a cyberculture." Turner, *From Counterculture to Cyberculture,* 148.

17. Hilton, "Basic Assumptions," in *Evolving Society,* 8.

18. Hilton, 4.

19. Hilton, 8–9. Williams first used the phrase "structure of feeling" in *Preface to Film,* cowritten with Michael Orrom (Film Drama, 1954).

20. Fredric Jameson, *The Political Unconscious: Narrative as a Socially Symbolic Act* (Routledge, 2002), 13.

21. Examples of the automation discourse include Karl Marx and Charles Babbage's work during the emergence of the industrial factory and discussions today from across the political spectrum, whether these be Andrew Yang's 2020 presidential campaign, those about OpenAI and ChatGPT that are everywhere as I write this, or more explicitly left-leaning formulations by Nick Srnicek, Alex Williams, and Aaron Bastani. For excellent histories of automation as they are connected to its contemporary mobilization, see Jason E. Smith, *Smart Machines and Service Work: Automation in an Age of Stagnation* (Reaktion Books, 2020); and Benanav, *Automation.* The tension between what was novel and what was continuous about automation in the post–World War II period that is complementary to Black Power discussions is well represented by Noble's and Braverman's approaches. Noble has shown that the emergence of the 1960s sense of automation was defined by the historically specific intersection of World War II, management's war against

labor, and a new scientific, technological, computer-based ideology of total control. However, although this historical intersection was new, Braverman notes that the introduction of mechanization and automation in the period was part of a long-term project in which scientific knowledge was incorporated into the production process by management in order to destroy the craft character of labor and weaken workers' control over production while leaving little else in its place except deepening class divisions. David Noble, *Forces of Production: A Social History of Industrial Automation* (Transaction, 2011); Harry Braverman, *Labor and Monopoly Capitalism: The Degradation of Work in the Twentieth Century* (Monthly Review, 1974).

22. Hilton, "The Bases of Cyberculture," in *Evolving Society,* 20.

23. See Jonathan Beller, *The World Computer: Derivative Conditions of World Capitalism* (Duke University Press, 2021); Ruha Benjamin, *Race After Technology: Abolitionist Tools for the New Jim Crow* (Polity, 2019); Simone Browne, *Dark Matters: On the Surveillance of Blackness* (Duke University Press, 2015); Wendy Chun, "Race and/as Technology," *Camera Obscura,* 24, no. 1 (2009); Kate Crawford, *Atlas of AI* (Yale University Press, 2021); Curtis Marez, *Farm Worker Futurism: Speculative Technologies of Resistance* (University of Minnesota Press, 2016); Alondra Nelson, "Introduction: Future Texts," *Social Text* 20, no. 2 (2002); Safiya Umoja Noble, *Algorithms of Oppression: How Search Engines Reinforce Racism* (NYU Press, 2018); and Astra Taylor, "The Automation Charade," *Logic(s),* no. 5 (August 1, 2018), logicmag.io/failure/the-automation-charade/.

24. Atanasoski and Vora, *Surrogate Humanity,* 29.

25. James Boggs, *Racism and the Class Struggle: Further Pages from a Black Worker's Notebook* (Monthly Review Press, 1970).

26. J. Boggs, "The City Is the Black Man's Land," in *Racism and the Class Struggle,* 40.

27. J. Boggs, 42.

28. The extent and effects of automation have been debated consistently since the nineteenth century. According to Thomas Sugrue, Detroit workers were hit hard by automation, although numbers vary. Between 1951 and 1953, Ford factories in Detroit lost 4,185 jobs to automation, and overall employment at the Ford River Rouge Plant declined from 85,000 jobs in 1945 to 30,000 in 1960. Thomas Sugrue, *The Origins of the Urban Crisis: Race and Inequality in Postwar Detroit* (Princeton University Press, 2014), 132–34.

29. J. Boggs, "The City Is the Black Man's Land," 43.

30. Smith, *Smart Machines.*

31. J. Boggs, "The City Is the Black Man's Land," 43.

32. J. Boggs, 50.

33. J. Boggs, 49.

34. Jackie Wang, *Carceral Capitalism* (semiotext(e), 2018), 56–57.

35. See, e.g., Angela Y. Davis, "Afro Images: Politics, Fashion, and Nostalgia," *Critical Inquiry* 21, no. 1 (1994).

36. Peniel E. Joseph, "Introduction: Toward a Historiography of the Black Power Movement," in *The Black Power Movement: Rethinking the Civil Rights–Black Power Era,* ed. Peniel E. Joseph (Routledge, 2006), 8; Ashley D. Farmer,

Remaking Black Power: How Black Women Transformed an Era (University of North Carolina Press, 2017).

37. Johnson examines the political struggles between nationalists, Marxists, and corporate-oriented elites that largely resulted in Black political incorporation into conventional American political structures in the 1970s. He resists the inclination to see Black Americans as an undifferentiated "corporate political entity" and instead argues that "the Black Power movement might best be understood as a historical debate over the character and address of post–Jim Crow race advancement projects rather than the projection of some common political will." Cedric Johnson, *Revolutionaries to Race Leaders: Black Power and the Making of African American Politics* (University of Minnesota Press, 2007), xxxviii, xxii.

38. See Addison Gayle Jr., ed., *The Black Aesthetic* (Doubleday, 1971); Larry Neal, "The Black Arts Movement," *Drama Review* 12, no. 4 (1968): 29–39; Amy Abugo Ongiri, *Spectacular Blackness: The Cultural Politics of the Black Power Movement and the Search for a Black Aesthetic* (University of Virginia Press, 2010); and Lisa Gail Collins and Margo Natalie Crawford, eds., *New Thoughts on the Black Arts Movement* (Rutgers University Press, 2006). In terms of its broader politics, Smethurst argues that "one could just as easily say that Black Power was the political wing of the Black Arts movement" instead of the other way around. James Smethurst, *The Black Arts Movement: Literary Nationalism in the 1960s and 1970s* (University of North Carolina Press, 2005), 14. On the media practices of the Black Panthers, see Sampada Aranke, *Death's Futurity: The Visual Life of Black Power* (Duke University Press, 2023).

39. Brandon M. Terry, "Requiem for a Dream: The Problem-Space of Black Power," in *To Shape a New World: Essays on the Political Philosophy of Martin Luther King Jr.*, ed. Tommie Shelby and Brandon M. Terry (Belknap Press, 2018).

40. To give only a few examples that could have been included: paralleling James and Grace Lee Boggs and the LRBW's Detroit-based political theory, Charles Denby, although critical of aspects of the Black Power movement, wrote in *Indignant Heart* (South End Press, 1978); *Workers Battle Automation* (pamphlet series, 1960–11); and in the pages of *News and Letters,* which he also edited, about the connections between factory automation, Southern segregation, and the continued influences of plantation slavery. One of the most important early figures of the Black Power movement was Robert Williams, who notes that one of the conditions that made a revolution in the United States possible was that a "highly industrialized and mechanized system is also a very sensitive one. The more machinery required to serve a community the greater the incidence of mechanical breakdown. The more dependent a community is on mechanization, the more important it is for the wheels of industry to perpetually turn smoothly." Robert F. Williams, "USA: The Potential of a Minority Revolution," *Crusader* 5, no. 4 (May/June 1964): 5. The Revolutionary Action Movement based their own theories of revolution in part on Williams's characterization of the technological limitations of White affluent society; see Revolutionary Action Movement, "World Black Revolution" (1966), *Viewpoint Magazine,* December 29, 2017, viewpointmag.com. Alternatively, Amiri Baraka in "Technology and Ethos,

Vol. 2 Book of Life" calls for an expansive reimagining of what technologies can do "spiritually" beyond being "extensions" of European genres of the human: Imamu Amiri Baraka, *Raise, Race, Rays, Raze: Essays Since 1965* (Random House, 1971), 155–57.

41. Raya Dunayevskaya, "Two Worlds," *News and Letters* 12, no. 1 (1967): 5, 7.

42. Noah Purifoy and Ted Michel, "The Art of Communication as a Creative Act," *Junk Art: 66 Signs of Neon* (66 Signs of Neon, 1966).

43. David Harvey, *The Condition of Postmodernity* (Wiley-Blackwell, 1991), 139.

44. Jennifer L. Morgan, "*Partus sequitur ventrem:* Law, Race, and Reproduction in Colonial Slavery," *Small Axe* 22, no. 1 (2018).

45. Cedric Robinson, *Black Marxism: The Making of the Black Radical Tradition* (University of North Carolina Press, 2000), 26.

46. See, e.g., Susan Koshy, Lisa Marie Cacho, Jodi A. Byrd, and Brian Jordan Jefferson, eds., *Colonial Racial Capitalism* (Duke University Press, 2022).

47. Destin Jenkins and Justin Leroy draw on Robinson's work and Saidiya Hartman's to effectively summarize the project of racial capitalist critique as one that "rethink[s] the past, present, and futures of capitalism. . . . It is also a methodological practice—a way of seeing—that asks practitioners to question the structuring idioms, themes, and subjects in the study of capitalism in the present. Finally, the analytic of racial capitalism also suggests a future-oriented political analysis that cautions against either a 'race first' or a 'class first' approach toward social justice." Destin Jenkins and Justin Leroy, eds., *Histories of Racial Capitalism* (Columbia University Press, 2021), 10.

48. Nikhil Pal Singh, "Black Marxism and the Antinomies of Racial Capitalism," in *After Marx: Literature, Theory, and Value in the Twenty-First Century*, ed. Colleen Lye and Christopher Nealon (Cambridge University Press, 2022).

49. Singh, 34.

50. See, e.g., Tithi Bhattacharya, ed., *Social Reproduction Theory: Remapping Class, Recentering Oppression* (Pluto Press, 2017). Nancy Fraser argues that it is especially this division between the "economic" and the "social," between "foreground" and "background," and the ways that the economic "cannibalizes" and appropriates its background conditions of possibility that is responsible for capitalism's "crisis tendency." It is especially when capitalism's growth drive and accumulation strategies throw reproduction into crisis that "boundary struggles" over these formal divisions and how society is reproduced take place, in turn creating "epoch transformations." Nancy Fraser, *Cannibal Capitalism. How Our System Is Devouring Democracy, Care, and the Planet and What We Can Do About It* (Verso, 2022), 58.

51. Jennifer Morgan argues that economic, technical, and mathematical rationality and histories of the slave trade are too easily partitioned; instead, the concepts of race and capitalist value as well as forms of inheritance and the social reproduction of capitalism emerged together as central features of modernity with the slave trade. Although Morgan focuses on a much different seventeenth-century moment, and thus requires a much different relationship to the archive than one does in the cybercultural era, asking how enslaved peoples related to and struggled against the emergence of

capitalism, the slave trade, and the plantation resonates with my attempt to elucidate Black Power's varied political responses to the cybercultural mode of production and reproduction. Jennifer L. Morgan, *Reckoning with Slavery: Gender, Kinship, and Capitalism in the Early Black Atlantic* (Duke University Press, 2021). On social reproduction in Jamaica in the context of contemporary forms of automation, see Beverley Mullings, "Caliban, Social Reproduction, and Our Future Yet to Come," *Geoforum* 118 (2021).

52. Angela Y. Davis, *Women, Race, and Class* (Vintage, 1983), 223.

53. Huey P. Newton, "The Technology Question," in *The New Huey P. Newton Reader*, ed. David Hilliard and Donald Weise (Seven Stories Press, 2019), 271.

54. Pamela M. Lee, *Chronophobia: On Time in the Art of the 1960s.* (MIT Press, 2004), 8.

55. Lee, 8.

56. Erik H. Erikson and Huey Newton, *In Search of Common Ground* (Norton, 1973), 16.

57. Erikson and Newton, 103.

58. Erikson and Newton, 32–33.

59. J. Boggs, *Pages,* 140.

60. J. Boggs, 140.

61. J. Boggs, 140–41.

62. Boggs and Boggs, *Revolution and Evolution,* 19.

63. Boggs and Boggs, 19.

1. "Remaining Awake Through a Great Revolution"

1. Martin Luther King Jr., "Remaining Awake Through a Great Revolution," sermon at the National Cathedral, Washington, D.C., March 31, 1968, kinginstitute.stanford.edu.

2. Ad Hoc Committee on the Triple Revolution (AHC), *The Triple Revolution* (Ad Hoc Committee on the Triple Revolution, 1964).

3. AHC, 6.

4. Martin Luther King Jr., "An Address by Dr. Martin Luther King Jr.," Chicago Freedom Festival, the Amphitheatre, Chicago, Illinois, March 12, 1966, www.crmvet.org/docs/6603_sclc_mlk_cfm.pdf.

5. AHC, *Triple Revolution,* 9.

6. Melinda Cooper, *Family Values: Between Neoliberalism and the New Social Conservatism* (Zone Books, 2017).

7. Tommie Shelby and Brandon M. Terry, "Introduction: Martin Luther King Jr. and Political Philosophy," in *To Shape a New World: Essays on the Political Philosophy of Martin Luther King Jr.,* ed. Tommie Shelby and Brandon M. Terry (Belknap Press, 2018), 8.

8. Shelby and Terry, 8–9. In this, they echo the work of scholars like Robin D. G. Kelley and Elizabeth Esch, for whom "the story of the shift from civil rights to 'Black Power' has been told so many times . . . that it has become a kind of common sense." In this "typology," "high expectations begot the civil rights movement; the movement's failure to achieve all its goals and to deal with urban poverty begot Black Power." Kelley and Esch argue that this narrative suffers because it is too narrowly domestic in its focus,

oversimplifies a South–North trajectory, and suppresses the socialist and Third World voices in the Black U.S. freedom struggle. Robin D. G. Kelley with Betsy Esch, *Freedom Dreams: The Black Radical Imagination* (Beacon Press, 2002), 62, 60.

9. Martin Luther King Jr., "Statement on Ending the Bus Boycott," speech, Montgomery, Alabama, December 20, 1956, kinginstitute.stanford.edu.

10. In this chapter, I also follow Shelby and Terry's claim about interpretations of King: that he is too often discussed through his "charismatic rhetorical persona" and in terms of "the celebration of, and reverence for, virtuosic oratory performance and oracular wisdom." Shelby and Terry, *To Shape,* 10. They instead call for a systematic engagement with King's published and written works, whether monographs or the transcripts of his speeches. This echoes arguments made by Hortense Spillers, who argues that although his oratorical style undoubtedly carried powerful emotional "stuff," his "political sermons" were meticulously written to carry his moral message, and "he did not leave the organization of his ideas to chance." Hortense J. Spillers, "Martin Luther King and the Style of the Black Sermon," *Black Scholar* 3, no. 1 (1971): 16.

11. Martin Luther King Jr., "Remaining Awake Through a Great Revolution," address at Morehouse College Commencement, Atlanta, Georgia, June 2, 1959, kinginstitute.stanford.edu.

12. The capitalizations, misspellings, and strikethroughs of *New York City* and *Tokyo* are as they appear in the written version of the speech, as transcribed by the Martin Luther King Jr. Research and Education Institute.

13. Harvey, *Condition of Postmodernity.*

14. King, "Remaining Awake," 1968 version.

15. "If the Machine Wants Our Jobs Let's Buy It," editorial, *Life,* August 14, 1964, 4.

16. Ad Hoc Committee, *Triple Revolution,* 5.

17. Ad Hoc Committee, 9.

18. National Commission on Technology, Automation, and Economic Progress, *Report Vol. 1: Technology and the American Economy* (U.S. Government Printing Office, 1966), 1.

19. I discuss Richard Nixon in more detail later. More generally, see Robert Lekachman, "The Automation Report," *Commentary,* May 1966, www.commentary.org.

20. Howard Brick, *Transcending Capitalism: Visions of a New Society in Modern American Thought* (Cornell University Press, 2006).

21. Cooper, *Family Values,* 50.

22. Frederick Pollock, *Automation: A Study of Its Economic and Social Consequences,* trans. W. O. Henderson and W. H. Chaloner (Frederick A. Praeger, 1957), 3.

23. Smith, *Smart Machines,* 22.

24. Howard Bowen and Garth L. Mangum, eds., *Automation and Economic Progress* (Prentice Hall, 1966), 1.

25. Benanav, *Automation,* 2.

26. AHC, *Triple Revolution,* 6.

27. National Commission, *Report Vol. 1,* 110.

28. AHC, *Triple Revolution,* 6.
29. AHC, 6–7.
30. Alternatively, see Michel Foucault, *The Birth of Biopolitics: Lectures at the Collège de France, 1978–1979,* ed. Michel Senellart and trans. Graham Burchell (Palgrave Macmillan, 2008); and Nick Srnicek and Alex Williams, *Inventing the Future: Postcapitalism and a World Without Work* (Verso, 2016).
31. Harvey, *Condition of Postmodernity,* 135.
32. Cooper, *Family Values,* 55.
33. AHC, *Triple Revolution,* 11.
34. AHC, 10.
35. Benanav, *Automation,* 12, 8, 22.
36. AHC, *Triple Revolution,* 10.
37. AHC, 9.
38. AHC, 13.
39. Editors, "The Triple Revolution," *Monthly Review* 16, no. 7 (1964).
40. Herbert Marcuse, *One-Dimensional Man: Studies in the Ideology of Advanced Industrial Society* (Beacon Press, 1991), 52.
41. On accelerationism, see Benjamin Noys, *Malign Velocities: Accelerationism and Capitalism* (Zero Books, 2014); and Steven Shaviro, *No Speed Limit: Three Essays on Accelerationism* (University of Minnesota Press, 2015).
42. Reinhart Koselleck, *Futures Past,* trans. Keith Tribe (Columbia University Press, 2004).
43. W. H. Ferry, "Education and the Triple Revolution," in *Voices of Crisis,* ed. Floyd W. Matson (Odyssey Press, 1967), 221.
44. King, "Address by Dr. Martin Luther King Jr."
45. Keeanga-Yamahtta Taylor, *Race for Profit: How Banks and the Real Estate Industry Undermined Black Homeownership* (University of North Carolina Press, 2019).
46. King, *All Labor Has Dignity,* ed. Michael K. Honey (Beacon Press, 2011), 174.
47. King, 174.
48. King, "Paul's Letter to American Christians," sermon at Dexter Avenue Baptist Church, Montgomery, Alabama, November 4, 1956, kinginstitute.stanford.edu.
49. Martin Luther King Jr., "Remarks for Negro Press Week," Montgomery, Alabama, February 10, 1958, kinginstitute.stanford.edu.
50. King, *All Labor Has Dignity,* 10.
51. King, 10.
52. Martin Luther King Jr., "Facing the Challenge of a New Age," address at NAACP Emancipation Day Rally, Atlanta, Georgia, January 1, 1957, kinginstitute.stanford.edu.
53. Martin Luther King Jr., *Where Do We Go from Here: Chaos or Community?* (Beacon Press, 2010), 196–97.
54. Quoted in Trevor Burnard and John Garrigus, eds., *The Plantation Machine: Atlantic Capitalism in French Saint-Domingue and British Jamaica* (University of Pennsylvania Press, 2016), 5.
55. C. L. R. James, *The Black Jacobins: Toussaint L'Ouverture and the San Do-*

mingo Revolution (Vintage, 1989), 85–86, 392; Louis Chude-Sokei, *The Sound of Culture: Diaspora and Black Technopoetics* (Wesleyan University Press, 2016), 36–39.

56. Aimé Césaire, *Discourse on Colonialism,* trans. Joan Pinkham (Monthly Review Press, 2000), 42–43.

57. King, *Where Do We Go from Here?,* 6–7.

58. Saidiya Hartman, "The Time of Slavery," *South Atlantic Quarterly* 101, no, 4 (2022): 759.

59. Citing Jesse Jackson, King remarked in Memphis the night before his assassination that the Sanitation Workers strike and the Poor People's Campaign were about "redistribute[ing] the pain" of racial capitalism. King, *All Labor Has Dignity,* 189.

60. Nikhil Pal Singh, "On Race, Violence, and 'So-Called Primitive Accumulation,'" in *Futures of Black Radicalism,* ed. Gaye Theresa Johnson and Alex Lubin (Verso, 2017), 51–52.

61. King, *All Labor Has Dignity,* 116.

62. King, 115–16.

63. King, 116.

64. King, 116.

65. King, *Where Do We Go from Here?,* 3.

66. King, 37.

67. Robert Allen, "Reassessing the Internal (Neo)colonialism Theory," *Black Scholar* 35, no. 1 (2005): 2–11. See also William Jalmer Kujala, "The Dialectics of Domestication: Domestic Colonialism and Internationalist Criticism in North America" (PhD diss., University of Alberta, 2023), era.library.ualberta.ca/items/37128575-cba0-4d10-8302-5f947b4c5990; and Cedric Johnson, "Huey P. Newton and the Last Days of the Black Colony," in *African American Political Thought: A Collected History,* ed. Melvin L. Rogers and Jack Turner (University of Chicago Press, 2021), 631–59.

68. Robert Allen, *Black Awakening in Capitalist America: An Analytic History* (Africa World Press, 1992), 228.

69. Although not Allen's position, King was seen by many Black radicals as playing a neocolonial mediator's role. See Allen, *Black Awakening,* 111–15.

70. For example, see King, "Remaining Awake," 1968 version, although his use of "World House" and "networks of mutuality" is widespread.

71. Allen, *Black Awakening,* 116.

72. King, *Where Do We Go from Here?,* 51.

73. King, 39.

74. King, *Why We Can't Wait* (Signet, 1964), 129–30.

75. King, *Where Do We Go from Here?,* 172.

76. Shatema Threadcraft and Brandon M. Terry, "Gender Trouble: Manhood, Inclusion, and Justice," in *To Shape a New World: Essays on the Political Philosophy of Martin Luther King Jr.,* ed. Tommie Shelby and Brandon M. Terry (Belknap Press, 2018).

77. King, *Where Do We Go from Here?,* 172.

78. Thomas F. Jackson, *From Civil Rights to Human Rights: Martin Luther King Jr. and the Struggle for Economic Justice* (University of Pennsylvania Press, 2007).

79. As quoted in Premilla Nadasen, *Welfare Warriors: The Welfare Rights Movement in the United States* (Routledge, 2005), 173.
80. Nadasen, 189.
81. Nadasen, xviii.
82. Nadasen, 165.
83. Nadasen, 166.
84. Saidiya Hartman, "The Belly of the World: A Note on Black Women's Labors," *Souls* 18, no. 1 (2016): 167.
85. Nadasen, *Welfare Warriors*, 122–23.
86. Allen, *Black Awakening*, 240, etc.
87. Sylvie Laurent, *King and the Other America: The Poor People's Campaign and the Quest for Economic Equality* (University of California Press 2018), 129.
88. King, *Where Do We Go from Here?*, 92.
89. King, 173.
90. See David P. Stein, "This Nation Has Never Honestly Dealt with the Question of a Peacetime Economy: Coretta Scott King and the Struggle for a Nonviolent Economy in the 1970s," *Souls* 18, no. 1 (2016): 80–105.
91. Martin Luther King Jr., "Statement by Dr. Martin Luther King Jr.," Atlanta, Georgia, December 4, 1967, www.crmvet.org/docs/6712_mlk_ppc-anc.pdf.
92. Michael K. Honey, *To the Promised Land: Martin Luther King and the Fight for Economic Justice* (Norton, 2018).
93. On the skepticism of many in the SCLC, see Laurent, *King and the Other America*, 12, 102.
94. King, *Where Do We Go from Here?*, 208–9.
95. Ruth Wilson Gilmore, "Abolition and the Problem of Innocence," in *Futures of Black Radicalism*, ed. Gaye Theresa Johnson and Alex Lubin (Verso, 2017), 238.
96. Honey, *To the Promised Land*.
97. John Wiebenson, "Planning and Using Resurrection City," in *Journal of the American Institute of Planners* 35, no. 6 (November 1969).
98. Laurent, *King and the Other America*, 196.
99. King, "Facing the Challenge of a New Age," 1957.
100. Gilmore, "Abolition," 227.
101. Gilmore, 231.
102. Gilmore, 238.

2. Initiating a New Plateau

1. Grace Lee Boggs with Scott Kurashige, *The Next American Revolution: Sustainable Activism for the Twenty-First Century* (University of California Press, 2011), 29.
2. James and Grace Lee Boggs, who are best known today as Detroit-based activists, had a unique relationship, especially for the United States in the mid-twentieth century. Grace Lee Boggs came from a middle-class Chinese American family; in the 1930s, she completed a PhD in philosophy at Bryn Mawr College. James Boggs was from rural Alabama and moved to Detroit during the Great Migration and remained an autoworker into the

1970s. They met through C. L. R. James and the Johnson-Forest Tendency, and helped found the journal *Correspondence.* See especially the essential biography of the first part of their lives: Ward, *In Love and Struggle;* and see also G. L. Boggs, *Living for Change.*

3. Boggs and Boggs, *Revolution and Evolution,* 182.

4. AHC, *Triple Revolution,* 11.

5. AHC, 6.

6. Joan Martinez-Alier, "Political Ecology, Distributional Conflicts, and Economic Incommensurability," *New Left Review* 1, no. 211 (1995): 70.

7. Martinez-Alier, 74.

8. *The American Revolution* as well as many of their essays from the 1960s were authored by James Boggs. However, some, like "The City Is the Black Man's Land," were attributed to both, and by the 1970s, they referred to all of this work as theirs. In this chapter, I will largely follow the latter convention, but when appropriate, I will also refer to a single author.

9. For example, Grace Lee Boggs made the first English-language translation of sections of Marx's *Economic and Philosophic Manuscripts of 1844.*

10. Jack Halberstam, *The Queer Art of Failure* (Duke University Press, 2011), 16.

11. Alexander Galloway, "Alexander R. Galloway—An Interview with McKenzie Wark," *b2o,* April 7, 2017, www.boundary2.org/2017/04/alexander-r-galloway-an-interview-with-mckenzie-wark.

12. J. Boggs, *Pages,* 97.

13. Bernard Stiegler, *For a New Critique of Political Economy* (Polity, 2010).

14. J. Boggs, *Pages,* 110.

15. J. Boggs, 334.

16. J. Boggs, *Racism and the Class Struggle,* 40.

17. Anna Lowenhaupt Tsing, *The Mushroom at the End of the World: On the Possibility of Life in Capitalist Ruins* (Princeton University Press, 2015).

18. J. Boggs, *Pages,* 113.

19. J. Boggs, 143.

20. J. Boggs, 86.

21. See especially Steve Wright, *Storming Heaven: Class Composition and Struggle in Italian Autonomist Marxism* (Pluto Press, 2017); Nicola Pizzolato, "Transnational Radicals: Labour Dissent and Political Activism in Detroit and Turin (1950–1970)," *IRSH* 56 (2011), 1–30; and Marcuse, "Introduction to the First Edition," *One-Dimensional Man.* For more on James Boggs's theory of automation, see Cedric Johnson, "James Boggs, the 'Outsiders,' and the Challenge of Postindustrial Society," *Souls,* 13, no. 3 (2011): 303–26; Patrick King, "Introduction to Boggs," *e-flux Journal,* no. 79 (February 2017), www.e-flux.com/journal/79/94671/introduction-to-boggs/; and Jason Smith, "Nowhere to Go: Automation, Then and Now," parts 1 and 2, *Brooklyn Rail,* March 2017 and April 2017.

22. J. Boggs, *Pages,* 86.

23. J. Boggs, 100.

24. See Cedric Robinson, *Black Marxism: The Making of the Black Radical Tradition* (University of North Carolina Press, 2020), 2. Frederick Douglass,

My Bondage and My Freedom (Yale University Press, 2014); and two books by David Roediger: *The Wages of Whiteness: Race and the Making of the American Working Class* (Verso, 2007) and *Class, Race, and Marxism* (Verso, 2019).

25. J. Boggs, *Pages,* 89.
26. J. Boggs, 87.
27. J. Boggs, 102.
28. J. Boggs, 102.
29. Karl Marx, *Grundrisse: Foundations of the Critique of Political Economy* (Penguin Classics, 1993), 706.
30. J. Boggs, *Pages,* 102.
31. J. Boggs, 102–3.
32. J. Boggs, 103.
33. J. Boggs, 137.
34. J. Boggs, *Pages,* 175–77.
35. Karl Marx and Frederick Engels, *Marx and Engels Collected Works,* vol. 24 (Lawrence and Wishart, 2010).
36. On this history, see Paul Thomas, *Marxism and Scientific Socialism: From Engels to Althusser* (Routledge, 2008).
37. Louis Althusser, *For Marx,* trans. Ben Brewster (Verso, 2005), 13.
38. On dialectical courterdiscourse, see George Ciccariello-Maher, *Decolonizing Dialectics* (Duke University Press, 2017).
39. For a complementary sense of blackness and racial capitalism, see Charisse Burden-Stelly, "Modern U.S. Racial Capitalism: Some Theoretical Insights," *Monthly Review,* July 1, 2020, monthlyreview.org/2020/07/01/modern-u-s-racial-capitalism/.
40. J. Boggs, *Pages,* 208.
41. J. Boggs, 187.
42. J. Boggs, 238.
43. Boggs and Boggs, *Revolution and Evolution,* 19.
44. J. Boggs, *Racism and the Class Struggle,* 50.
45. Boggs and Boggs, *Revolution and Evolution,* 160.
46. See, e.g., J. Boggs, "Community Building: An Idea Whose Time Has Come," in *Pages,* 331–40.
47. See G. L. Boggs, *Next American Revolution,* chap. 4.
48. Rebecca Solnit, "Detroit Arcadia: Exploring the Post-American Landscape," *Harper's Magazine,* July 2007.
49. Boggs and Boggs, *Revolution and Evolution,* 200.
50. J. Boggs, *Pages,* 234.
51. J. Boggs, 174.
52. J. Boggs, 187.
53. Alongside this, and especially relevant to the Boggses, Moore argues that "Cheap Nature, as an accumulation strategy, works by reducing the value composition—but increasing the technical composition—of capital as a whole; by opening new opportunities for the investment; and, in its qualitative dimension, by allowing technologies and new kinds of nature to transform extant structures of capital accumulation and world power." Jason Moore, *Capitalism in the Web of Life: Ecology and the Accumulation of Capital* (Verso, 2015), 53.

54. Deborah Fitzgerald argues that the "fundamental feature of twentieth-century agriculture" was the emergence of an "industrial logic or ideal" (3) in which, beginning in the 1920s, "the ultimate meaning of mechanization" was "displacing farmers themselves with machines [as] the only rational outcome of mechanization" (92–93). Deborah Fitzgerald, *Every Farm a Factory: The Industrial Ideal in American Agriculture* (Yale University Press, 2010). On farm mechanization's impacts on Black agricultural workers in the 1940s, see "Mechanization in Agriculture (1941–1946)," in *A Hammer in Their Hands: A Documentary History of Technology and the African American Experience,* ed. Carroll Pursell (MIT Press, 2005), 260–64.

55. On the complex relationship between Red Power and Black Power written during the period, see Vine Deloria Jr., *Custer Died for Your Sins: An Indian Manifesto* (University of Oklahoma Press, 1988). For contemporary analyses, see especially Nick Estes, *Our History Is the Future* (Verso, 2019); Kent Blansett, *A Journey to Freedom: Richard Oakes, Alcatraz, and the Red Power Movement* (Yale University Press, 2018); and Roxanne Dunbar-Ortiz, *An Indigenous Peoples' History of the United States* (Beacon Press, 2014). For the history of settler colonialism and the "political economy of plunder" during the eighteenth and nineteenth centuries in the Anishinaabe land that is now Michigan, see Michael John Witgen, *Seeing Red: Indigenous Land, American Expansion, and the Political Economy of Plunder in North America* (Omohundro Institute of Early American History and Culture, 2022).

56. J. Boggs, *Racism and the Class Struggle,* 40.

57. Boggs and Boggs, *Revolution and Evolution,* 148–49. For an analysis of settler colonialism, racial capitalism, and land, see Glen Sean Coulthard, *Red Skin, White Masks: Rejecting the Colonial Politics of Recognition* (University of Minnesota Press, 2014); and Susan Koshy, Lisa Marie Cacho, Jodi A. Byrd, and Brian Jordan Jefferson, eds., *Colonial Racial Capitalism* (Duke University Press, 2022).

58. J. Boggs, *Pages,* 188.

59. "A Unique Stage in Human Development" is the title of chapter 8 of Boggs and Boggs, *Revolution and Evolution.*

60. See, e.g., Greg Castillo, "Counterculture Terroir: California's Hippie Environmental Zone," and Simon Sadler, "Mandalas or Raised Fists? Hippie Holism, Panther Totality, and Another Modernism," in *Hippie Modernism: The Struggle for Utopia,* ed. Andrew Blauvelt (Walker Art Center, 2015). On feedback and ecological thought in the 1970s, see Daniel Belgrad, *The Culture of Feedback: Ecological Thinking in Seventies America* (University of Chicago Press, 2019).

61. Françoise Vergès, "Racial Capitalocene," in *Futures of Black Radicalism,* ed. G. T. Johnson and A. Lubin (Verso, 2017).

62. Vergès, "Racial Capitalocene," 80.

63. Interestingly, Deloria, in a discussion on the relationship between Red Power and Black Power, notes that, like "Indian people," "to survive, blacks must have a homeland where they can withdraw, drop the facade of integration, and be themselves," although he does not say where that should be. Deloria, *Custer Died for Your Sins,* 194.

64. Bill Mullen, *Afro-Orientalism* (University of Minnesota Press, 2004), 105. For a different but complementary discussion of this, see also Sara

Safransky, *The City After Property: Abandonment and Repair in Postindustrial Detroit* (Duke University Press, 2023), esp. chap 2.

65. Boggs and Boggs, *Revolution and Evolution,* 193.

66. Tsing, *Mushroom,* viii. See William Cronon, *Nature's Metropolis* (Norton, 1992).

67. J. Boggs, *Pages,* 208.

68. J. Boggs, 108.

69. Grace Lee Boggs, "Neither White nor Black—But Revolutionary," *Correspondence,* June 1963, as quoted in Ward, *In Love and Struggle.*

70. G. L. Boggs, *Next American Revolution,* 70–71.

71. Boggs and Boggs, *Revolution and Evolution,* 140.

72. Boggs and Boggs, 208.

73. Mullen, *Afro-Orientalism,* 116.

74. Boggs and Boggs, *Revolution and Evolution,* 193.

75. J. Boggs, *Pages,* 108, 115.

76. Matthias Schmelzer, *The Hegemony of Growth: The OECD and the Making of the Economic Growth Paradigm* (Cambridge University Press, 2016).

77. Jeremy L. Caradonna et al., "A Call to Look Past an Ecomodernist Manifesto: A Degrowth Critique," *Resilience,* May 6, 2015, www.resilience.org/stories/2015-05-06/a-degrowth-response-to-an-ecomodernist-manifesto.

78. See, e.g., Ekaterina Chertkovskaya et al., eds., *Towards a Political Economy of Degrowth* (Rowman & Littlefield, 2019); and Matthias Schmelzer, Aaron Vansintjan, and Andrea Vetter, *The Future Is Degrowth: A Guide to a World Beyond Capitalism* (Verso, 2022).

79. J. Boggs, *Pages,* 213.

80. Boggs and Boggs, *Revolution and Evolution,* 236.

81. See, e.g., James O'Toole and Others, *Work in America: Report of a Special Task Force to the Secretary of Health, Education, and Welfare* (Department of Health, Education, and Welfare, 1972).

82. J. Boggs, *Pages,* 163.

83. Hannah Arendt, *The Human Condition* (University of Chicago Press, 2018).

84. See the "Panel Discussion" that follows Arendt's presentation "On the Human Condition," in Hilton, *Evolving Society,* 221–53.

85. J. Boggs, *Pages,* 213. On a Hegelian ethics of activity that resonates with the Boggses, see Fredric Jameson, *The Hegel Variations* (Verso, 2010), chap. 6.

86. On this importance of working-class movements to degrowth politics, see especially Stefania Barca, "The Labor(s) of Degrowth," *Capitalism Nature Socialism* 30, no. 2 (2017): 207–16.

87. J. K. Gibson-Graham, *A Postcapitalist Politics* (University of Minnesota Press, 2006), 88, 91.

88. Boggs and Boggs, *Revolution and Evolution,* 214.

89. J. Boggs, *Pages,* 305.

90. Sylvia Wynter, "Race and Our Biocentric Belief System: An Interview with Sylvia Wynter," in *Black Education: A Transformative Research and Action Agenda for the New Century,* ed. Joyce E. King (American Educational Research Association, 2005), 363.

91. Wynter, 363. Wynter's thought, which ranges from the disciplines of

history to cognitive science, covers different terrain than James and Grace Lee Boggs's Detroit-based movement activism and theorizing. However, like dialectical humanism, Wynter's project is an attempt to understand the historical development of the specific overrepresented genre of the human she calls Man in order to begin to open up a space for a reconceptualization of the human. Wynter defines the current genre of Man as Man2, or the rational figure of accumulation *homo oeconomicus*, a post-Enlightenment, bioeconomic genre of the human that developed in the context of colonialism and the biological sciences. See especially Katherine McKittrick, ed., *Sylvia Wynter: On Being Human as Praxis* (Duke University Press, 2015).

92. Boggs and Boggs, *Revolution and Evolution*, 168.

93. Nicholas Mirzoeff, "Below the Water: Black Lives Matter and Revolutionary Time," *e-flux Journal*, no. 79 (February 2017), www.e-flux.com/journal/79/94164/below-the-water-black-lives-matter-and-revolutionary-time/.

94. Boggs and Boggs, *Revolution and Evolution*, 168.

3. Value in the Materialist World

1. Noah Purifoy, *High Desert* (Steidl, 2015), 15. For more information on the Noah Purifoy Outdoor Museum, see Franklin Sirmans and Yael Lipschutz, eds., *Noah Purifoy: Junk Dada* (Los Angeles County Museum of Art, 2015).

2. Yael Lipschutz, "Mojave in Me: Noah Purifoy on My Mind," *Brooklyn Rail*, October 2014. For more on *Bessemer Steel*, see Brian Bartell, "Noah Purifoy's Aesthetics for the Racial Capitalocene: Reading *66 Signs of Neon*," *Cultural Critique* 112 (2021): 24–58.

3. Robert Smithson, *Robert Smithson: The Collected Writings*, ed. Jack Flam (University of California Press, 1996), 106.

4. Assemblage artist Thornton Dial also lived in Bessemer, where he was employed as a metalworker.

5. Kellie Jones, "The Story of the Philosopher-Artist of L.A.: On the Life of Noah Purifoy, Keeper of the Watts Towers," *Literary Hub*, May 25, 2017, lithub.com/the-story-of-the-philosopher-artist-of-l-a/.

6. Noah Purifoy and Ted Michel, "The Art of Communication as a Creative Act," *Junk Art: 66 Signs of Neon* (66 Signs of Neon, 1966).

7. Purifoy, *High Desert*, 17.

8. See Kellie Jones, *South of Pico: African American Artists in Los Angeles in the 1960s and 1970s* (Duke University Press, 2017), 81–84; and Yael Lipschutz, "*66 Signs of Neon* and the Transformative Art of Noah Purifoy," in *L.A. Object and David Hammons Body Prints*, ed. Connie Rogers Tilton and Lindsay Charlwood (Tilton Gallery, 2011), 217. On the *66 Signs of Neon* sculptures being thrown away, see Noah Purifoy, "African American Artists of Los Angeles: Noah Purifoy," interview by Karen Anne Mason, UCLA Library for Oral History Research, Los Angeles, California, 1992, 97, oralhistory.library.ucla.edu/.

9. Purifoy and Michel, "Art of Communication."

10. Noah Purifoy, interview by Karen Anne Mason, "African American Artists of Los Angeles" (University of California at Los Angeles Oral History Progam, 1992), 42.

11. The *Junk Art: 66 Signs of Neon* catalog is online at eastofborneo.org/archives/the-oddity-of-found-things-66-signs-of-neon-catalog/.

12. Jussi Parikka, *A Geology of Media* (University of Minnesota Press, 2015).

13. With this term, Denning argues that there needs be a move away from critical terms like "bare life" and, especially, "wasted life" that figure people as akin to "garbage" (96). In doing so, he emphasizes that "capitalism begins not with the offer of work, but with the imperative to earn a living" and that capitalism must necessarily be conceptualized in terms of the dialectical relationship between so-call formal and informal economies that call into question other normative concepts like "unemployment" and "employment" (80). Denning's essay is in part a history of the normalization of these categories, of which the Keynesian era's focus on a full-employment economy is one important moment (84–86). Michael Denning, "Wageless Life," *New Left Review* 66 (November/December 2010).

14. Jones, *South of Pico,* 136.

15. See Sylvia Wynter, "Unsettling the Coloniality of Being/Power/Truth/Freedom: Towards the Human, After Man, Its Overrepresentation—An Argument," *CR: The New Centennial Review* 3, no. 2 (2013).

16. Vergès, "Racial Capitalocene," 73. The remark on constant capital is from Robinson, *Black Marxism,* 309.

17. Purifoy and Michel, "Art of Communication."

18. Roderick A. Ferguson, "Purifoy: The Shit, the World, Their Remaking," *South Atlantic Quarterly* 119, no. 3 (2020); and see Daniel Widener, *Black Arts West: Culture and Struggle in Postwar Los Angeles* (Duke University Press, 2010), 57, esp. "Studios in the Street," on post–Watts Rebellion Black Art.

19. The date of the preface: Governor's Commission on the Los Angeles Riots, *Violence in the City—An End or a Beginning?* (State of California, 1965), i–iii (hereafter McCone Report).

20. See, e.g., Robert Fogelson, "White on Black: A Critique of the McCone Commission Report on the Los Angeles Riots," *Political Science Quarterly* 82, no. 3 (1967): 337–67.

21. McCone Report, 85–86.

22. Purifoy, interview by Mason, 68.

23. On the response to the McCone Report, see, e.g., "McCone Report Unnoticed in Watts" (Los Angeles, California) *Afro-American,* December 18, 1965; "McCone Report Censured," *Los Angeles Sentinel,* December 22, 1966; "Civil Rights Group Assails McCone Report as 'Bitter Disappointment,'" *Los Angeles Times,* January 23, 1966.

24. Widener, *Black Arts West,* 99–100.

25. Bayard Rustin, "The Watts 'Manifesto' and the McCone Report," *Commentary* 41, no. 3 (1966): 30.

26. Edward Soja, "Los Angeles, 1965–1992: From Crisis-Generated Restructuring to Restructuring-Generated Crisis," in *The City: Los Angeles and Urban Theory at the End of the Twentieth Century,* ed. Allen J. Scott and Edward Soja (University of California Press, 1996), 431.

27. McCone Report, 5.

28. Rustin, "Watts 'Manifesto,'" 29.
29. McCone Report, 9.
30. Ferguson, "Purifoy," 449–50.
31. McCone Report, 43.
32. McCone Report, 3.
33. McCone Report, 8.
34. In the mid-1960s, a large percentage of Watts residents had migrated from the South. The McCone Report notes this primarily through crime statistics: "Of the adults arrested, 2,057 (out of 3,438) were born in 16 southern states" (24). Soja notes that six hundred thousand Black people migrated to Los Angeles, primarily from the South, between 1942 and 1965, as well as an equal number of largely poor White Southerners. Soja, "Los Angeles, 1965–1992," 430.
35. McCone Report, 81–82.
36. J. Boggs, *Pages,* 188.
37. Marcuse, *One-Dimensional Man,* 17.
38. Marcuse, xlviii.
39. Lee, *Chronophobia,* 28.
40. For more on Marcuse in relation to the Black Power movement and James and Grace Lee Boggs, see Cedric Johnson, "The 'Negro Revolution' and Cold War America," in *Revolutionaries to Race Leaders.*
41. Herbert Marcuse, *Eros and Civilization: A Philosophical Inquiry into Freud* (Beacon Press, 1966), xii–xiii.
42. Marcuse, 134.
43. Kellie Jones, "Black West, Thoughts on Art in Los Angeles," in *New Thoughts on the Black Arts Movement,* ed. Lisa Gail Collins and Margo Natalie Crawford (Rutgers University Press, 2006), 49.
44. Purifoy, "African American Artists," 16.
45. Purifoy and Michel, "Art of Communication."
46. Purifoy and Michel.
47. Ferguson, "Purifoy," 455.
48. Purifoy, "African American Artists," 77–78.
49. Purifoy and Michel, "Art of Communication."
50. On waste and junk as the "history" and "process" of "sorting and categorizing forms of waste" more generally, see Gillian Whiteley, *Junk: Art and the Politics of Trash* (I. B. Tauris, 2011), 16.
51. Purifoy, "African American Artists," 82.
52. Lipschutz, "*66 Signs of Neon* and the Transformative Art of Noah Purifoy," 232.
53. Marshall McLuhan, *Understanding Media: The Extensions of Man* (MIT Press, 1994), 7.
54. McLuhan, 8.
55. McLuhan, 9.
56. Smithson, *Collected Writings,* 100–101, 104. See also Parikka, *Geology of Media,* esp. chap 1, "Materiality: Grounds of Media and Culture."
57. Smithson, *Collected Writings,* 106.

58. Purifoy and Michel, "Art of Communication."

59. Kathryn Yusoff, *A Billion Black Anthropocenes or None* (University of Minnesota Press, 2018), 16.

60. Purifoy and Michel, "Art of Communication."

61. Jones, *South of Pico,* 71.

62. Norbert Wiener, *Cybernetics, Or Control and Communication in the Animal and the Machine* (MIT Press, 1985).

63. Wiener, *Cybernetics,* 16.

64. Wiener, 16.

65. For another sense of cybernetics as a politics of control, see Tiqqun, *The Cybernetic Hypothesis* (semiotext(e), 2020).

66. Guy Debord, "The Decline and Fall of the Spectacle-Commodity Economy," trans. Ken Knabb, *Internationale Situationniste,* no. 10 (March 1966), libcom.org/article/decline-and-fall-spectacle-commodity-economy-guy-debord.

67. Purifoy and Michel, "Art of Communication."

68. See, e.g., Luc Boltanski and Eve Chiapello, *The New Spirit of Capitalism,* trans. Gregory Elliot (Verso, 2007).

69. As quoted in Jones, *South of Pico,* 70.

70. C. L. R. James, Grace C. Lee, and Pierre Chaulieu, *Facing Reality* (Bewick Editions, 1974), 105. See also its section on why worker control of automation was both logical and necessary: "Automation and the Total Crisis."

71. Christopher Taylor, "The Refusal of Work: From the Postemancipation Caribbean to Post-Fordist Empire," *Small Axe* 18, no. 2 (2014): 1–17.

72. Taylor, 6, 4.

73. Taylor, 15.

74. Sylvia Wynter, *Black Metamorphosis: New Natives in a New World,* 616; undated and unpublished manuscript, Schomburg Center for Research in Black Culture, New York Public Library, New York. On *Black Metamorphosis,* see *Small Axe* 20, no. 1 (2016), particularly its essays by David Scott, Aaron Kamugisha, Greg Thomas, Nijah Cunningham, Katherine McKittrick, and Demetrius L. Eudell, whose "From Mode of Production to Mode of Auto-Institution: Sylvia Wynter's Black Metamorphosis of the Labor Question" drew my attention to the remark on labor. See also Yusoff, *Billion Black Anthropocenes.*

75. J. Boggs, *Pages,* 110.

76. Jones, "Black West," 48.

77. Purifoy and Michel, "Art of Communication."

78. Richard Cándida Smith, "Learning from Watts Towers: Assemblage and Community-Based Art in California," *Oral History* 37, no. 2 (209): 56.

4. Material Imperialism and the IBM Machine

1. Paule Marshall, *The Chosen Place, the Timeless People* (Harcourt, Brace, & World, 1969), 378.

2. Marshall, 380.

3. Eden Medina, *Cybernetic Revolutionaries: Technology and Politics in Allende's Chile* (MIT Press, 2014).

4. See, e.g., Robin D. G. Kelley and Betsy Esch, "Black Life Mao: Red China and Black Revolution," *Souls* 1, no. 4 (1999). For two quite different examples, see Stokely Carmichael, *Black Power and The Third World* (Third World Information Service, 1967); and the proposal for the Lumumba-Zapata College at the University of California, San Diego, at library.ucsd.edu/speccoll/DigitalArchives/ld781_s2-186-1969/. For the history of the Caribbean influence on Black radicalism in the United States, see Winston James, *Holding Aloft the Banner of Ethiopia: Caribbean Radicalism in Early Twentieth Century* (Verso, 1999).

5. J. Boggs, *Pages*, 187.

6. Edward Brathwaite, "Rehabilitations: West Indian History and Society in the Art of Paule Marshall's Novel," *Caribbean Studies* 10, no 2 (1970): 126.

7. Brathwaite, 125.

8. Paule Marshall, *Triangular Road* (Basic Books, 2009), 126.

9. Hortense Spillers, "*The Chosen Place, the Timeless People:* Some Figurations on the New World," in *Conjuring: Black Women, Fiction, and Literary Tradition*, ed. Marjorie Pryse and Hortense Spillers (Indiana University Press, 1985), 155.

10. Marshall, *Chosen Place*, 107.

11. Ann Laura Stoler, "Imperial Debris: Reflections on Ruins and Ruination," *Cultural Anthropology* 23, no. 2 (2008): 194.

12. Fredric Jameson, *The Antinomies of Realism* (Verso, 2015).

13. Saidiya Hartman, "The Time of Slavery," *South Atlantic Quarterly* 101, no. 4 (2022): 759.

14. Marshall, *Chosen Place*, 12.

15. Marshall, 14.

16. Gilbert Rist, *The History of Development: From Western Origins to Global Faith*, trans. Patrick Camiller (Zed Books, 2019).

17. See Rist, chap. 4, "The Invention of Development."

18. Michael Adas, *Dominance by Design: Technological Imperatives and America's Civilizing Mission* (Belknap Press, 2006), 243.

19. W. W. Rostow, *The Stages of Economic Growth: A Non-Communist Manifesto* (Cambridge University Press, 1960).

20. Marshall, *Chosen Place*, 14.

21. See, e.g., Janae Davis, Alex A. Moulton, Levi Van Sant, and Brian Williams, "Anthropocene, Capitalocene, . . . Plantationocene? A Manifesto for Ecological Justice in an Age of Global Crises," *Geography Compass* 13, no. 5 (2019); Donna Haraway, "Anthropocene, Capitalocene, Plantationocene, Chthulucene: Making Kin," *Environmental Humanities* 6, no. 1 (2015); Christian Høgsbjerg, "The Very Valley of the Shadow of Death: C. L. R. James on Capitalism and Environmental Destruction," *Radical History Review* 145 (2023); and Anna Lowenhaupt Tsing, "On Nonscalability: The Living World Is Not Amenable to Precision-Nested Scales," *Common Knowledge* 18, no. 3 (2012).

22. Justin Haynes, "Ghosts in the Posthuman Machine: Prostheses and Performance in *The Chosen Place, the Timeless People*," *Anthurium: A Caribbean Studies Journal* 14, no. 1 (2017): 1.

23. Haynes, 1.

24. Marshall, *Chosen Place,* 154.
25. Stephanie Smallwood, *Saltwater Slavery: A Middle Passage from Africa to American Diaspora* (Harvard University Press, 2007), 35–36.
26. Robinson, *Black Marxism,* 309.
27. C. L. R. James, *Black Jacobins,* 85–86.
28. Marshall, *Chosen Place,* 197.
29. Marshall, 56.
30. Marshall, 58.
31. Marshall, 185.
32. Lloyd Best and Kari Polanyi Levitt, *Essays on the Theory of Plantation Economy: A Historical and Institutional Approach to Caribbean Economic Development* (University of the West Indies Press, 2009), 6.
33. Best and Polanyi Levitt, 21.
34. Marshall, *Chosen Place,* 193.
35. Today, the place that Cassia House is based on, the Hotel Atlantis, is a luxury resort in Barbados.
36. Marshall, *Chosen Place,* 107.
37. Marshall, 401.
38. Marshall, 401.
39. Jane Bennett, "The Force of Things: Steps Toward an Ecology of Matter," *Political Theory* 32, no. 3 (2004).
40. Frederic Jameson, *Singular Modernity* (Verso, 2002), 40.
41. Lizabeth Paravisini-Gebert, "Paule Marshall's *The Chosen Place, the Timeless People:* Untenable Sisterhoods," *Journal of Caribbean Studies,* 1997, 58.
42. Marshall, *Chosen Place,* 389.
43. Marshall, 157.
44. Michel Foucault, *Society Must Be Defended: Lectures at the Collège de France, 1975–1976,* trans. David Macey (Picador, 2003), 245.
45. Marshall, *Chosen Place,* 243.
46. Marshall, 374.
47. Jonathan Beller, *The Message Is Murder: Substrates of Computational Capital* (Pluto Press, 2018), 5–6.
48. Marshall McLuhan, *Understanding Media: The Extensions of Man* (MIT Press, 1994), 9.
49. Paule Marshall, *Conversations with Paule Marshall,* ed. James C. Hall and Heather Hathaway (University of Mississippi Press, 2010), 24.
50. Marshall, *Chosen Place,* 17.
51. Marshall, 378.
52. Marshall, 380.
53. Marshall, 13.
54. Eldridge Cleaver, "Convalescence," in *Soul on Ice* (Delta, 1992), 222.
55. Cleaver, 233.
56. Hayles, *How We Became Posthuman,* 3.
57. Hayles, 4.
58. Alexander Weheliye, "Feenin': Posthuman Voices in Contemporary Black Popular Music," *Social Text* 20, no. 2 (2002): 23.
59. Weheliye, 26.

60. Louis Chude-Sokei, *The Sound of Culture: Diaspora and Black Technopoetics* (Wesleyan University Press, 2016). See esp. chap. 2, "Humanizing the Machine."

61. Norbert Wiener, *The Human Use of Human Beings: Cybernetics and Society* (Free Association Books, 1989), 162.

62. Wiener, *Human Use of Human Beings,* 162.

63. Wiener, 48.

64. See Belgrad, *Culture of Feedback.* Belgrad discusses the revisions that Wiener made to the sections of *The Human Use of Human Beings* on pp. 24–31.

65. Hayles, *How We Became Posthuman,* 24.

66. Marshall, *Chosen Place,* 467. On the shift from cybernetics to information, see Kline, *Cybernetics Moment.*

67. See Katherine McKittrick, *Demonic Grounds: Black Women and the Cartographies of Struggle* (University of Minnesota Press, 2006); and Katherine McKittrick, ed., *Sylvia Wynter: On Being Human as Praxis* (Duke University Press, 2015).

68. McKittrick, *Demonic Grounds,* 127–30.

69. McKittrick, 4–5.

70. Marshall, *Chosen Place,* 462–63.

71. Belinda Edmondson, *Making Men: Gender, Literary Authority, and Women's Writing in Caribbean Narrative* (Duke University Press, 1999), 5.

72. Marshall, *Chosen Place,* 6.

73. Marshall, 468.

5. "DRUM Would Like to Welcome You to the Plantation"

1. Housekeeping Assignments Notebook, Box 1, Folder 31, Detroit Revolutionary Movements Records, LR000874, Walter P. Reuther Library, Wayne State University, Detroit, Mich.

2. On this see, Marian Kramer's contribution to *Detroit Lives,* ed. Robert Mast (Temple University Press, 1994), 103–5; and Ursula McTaggart, *Guerrillas in the Industrial Jungle: Radicalism's Primitive and Industrial Rhetoric* (SUNY Press, 2012), 80–93. On gender divisions in Newsreel and the League, as well as their effect on both the content and production of *Finally Got the News,* see Chris Robé, "Detroit Rising: The League of Revolutionary Black Workers, Newsreel, and the Making of *Finally Got the News,*" *Film History* 28, no. 4 (2016).

3. Dan Georgakas and Marvin Surkin, *Detroit: I Do Mind Dying: A Study in Urban Revolution* (Haymarket, 2012), 1; *Finally Got the News,* dir. Stewart Bird et al. (Black Star Productions, 1970).

4. *Leviathan* 2, no. 2 (June 1970) in DRUM, Handbills, Booklets, Box 1, Folder 6, Detroit Revolutionary Movements Records, LR000874, Walter P. Reuther Library.

5. Throughout, I use lowercase *drum* when referring to the paper because it appears in lowercase on the masthead; and I use all-caps DRUM when referring to the Dodge Revolutionary Union Movement.

6. Ernie Allen, "Dying from the Inside: The Decline of the League of Revolutionary Black Workers," in *They Should Have Served that Cup of Coffee:*

7 Radicals Remember the '60s, ed. Dick Cluster (South End Press, 1990); Dan Georgakas, "*Finally Got the News:* The Making of a Radical Film," *Cinéaste* 5, no. 4 (1973).

7. Fredric Jameson, *Postmodernism, or, The Cultural Logic of Late Capitalism* (Verso, 1991), 414–15.

8. Peter Linebaugh and Bruno Ramirez, "Crisis in the Auto Sector," *Zerowork* 1 (1975): 83, zerowork.org.

9. Jordan T. Camp, "Challenging the Terms of Order: Representations of the Detroit Rebellions, 1967–1968," *Kalfou* 2, no. 1 (2015); Errol A. Henderson, *The Revolution Will Not Be Theorized: Cultural Revolution in the Black Power Era* (SUNY Press, 2019); Fred Moten, *In the Break: The Aesthetics of the Black Radical Tradition* (University of Minnesota Press, 2003).

10. Technically, Watson was referring here to activities around *Inner-City Voice,* like its coffee house and school. However, that the League was only a "loose conglomeration" was one of the critiques against it made by General Baker and Charles Wooten after the split with Watson, Ken Cockrel, and Mike Hamlin. In the wake of this, Baker, Wooten, and others sought a "strategic retreat" and a refocus on worker organizing and establishing the "dictatorship of the proletariat." John Watson, "The Black Editor: An Interview," *Radical America* 2, no. 4 (July–August 1968); LRBW (League of Revolutionary Black Workers), General Meeting, January 2, 1972, Box 1, Folder 25, Detroit Revolutionary Movements Records, LR000874, Walter P. Reuther Library.

11. For more on Detroit and the historical context of the League's emergence, see Georgakas and Surkin, *Detroit;* and James A. Geschwender, *Class, Race, and Worker Insurgency: The League of Revolutionary Black Workers* (Cambridge University Press, 1977).

12. Georgakas and Surkin, *Detroit,* 70.

13. See also Angela Davis, *An Autobiography* (Haymarket, 2023), "Part Four: Flames," on Davis's Marxist reception of Stokely Carmichael from the same moment.

14. Watson, "Black Editor."

15. Ken Cockrel, *League of Revolutionary Black Workers on Repression Speech,* 11–12, 1970, Box 1, Folder 92, American Left Ephemera Collection, Archives Service Center, University of Pittsburgh.

16. McTaggart, *Guerrillas,* 65.

17. Cockrel, *League of Revolutionary Black Workers on Repression Speech,* 12.

18. Cockrel, 13.

19. *Leviathan* (1970), 13.

20. All cited issues of *drum* are collected at DRUM, 1968–71, undated (newsletter), Box 17, Folder 4–8, Detroit Revolutionary Movements Records, LR000874, Walter P. Reuther Library.

21. See Georgakas and Surkin, *Detroit.* See also documents like the "Anti-ELRUM" letter in the General Baker papers: Employee's Committee on Human Equality (Anti-ELRUM), Box 1, Folder 16, Detroit Revolutionary Movements Records, LR000874, Walter P. Reuther Library.

22. Multiple documents related to this can be found in the Ronald Glotta collection of the Detroit Revolutionary Movement Papers, Subseries A: Chrys-

ler Eldon (ELRUM) and Other Revolutionary Union Movements, 1967–74, Detroit Revolutionary Movements Records, LR000874, Walter P. Reuther Library.

23. Jonathan Flatley, "How a Revolutionary Counter-Mood Is Made," *New Literary History,* 43, no. 3 (2012).

24. See esp. Browne, *Dark Matters.*

25. The available digital copy states 4, followed by a space. This is likely a problem with the scan; James Geschwender says it was 49. Geschwender, *Class, Race, and Worker Insurgency,* 90.

26. John Watson, "To the Point of Production: An Interview with John Watson," (Detroit, Michigan) *Fifth Estate,* May 1969, riseupdetroit.org/to-the-point-of-production-an-interview-with-john-watson-1969/.

27. *Leviathan* (1970), 35.

28. Smith, *Smart Machines,* 22.

29. *Leviathan* (1970), 35.

30. Hamlin in *Leviathan* (1970), 13.

31. György Lukács, "Reification and the Consciousness of the Proletariat," in *History and Class Consciousness: Studies in Marxist Dialectics,* trans. Rodney Livingstone (MIT Press, 1971), 84, 90.

32. Lukács, 88.

33. Lukács, 89–90.

34. Hamlin in *Leviathan* (1970), 13.

35. Marx, *Grundrisse,* 699, 704.

36. Marx, 704–5. In the "Fragment" in *Grundrisse,* Marx argues that with fixed capital, as the full development of command over labor and nature, the valorization process extends into society as whole. Yet to the extent that capital continues to posit labor time as the "sole measure and source of wealth," it is contradictory in that automated machinery simultaneously demands greater intensity and "superfluous" work from living labor while also reducing necessary labor time "to a minimum" (706). In the short term, workers continue to be linkages in the automated machine, which mediates their metabolic relationship to nature. The development of the contradictions of fixed capital, Marx also notes, opens up the possibility that reduced necessary labor time would ultimately result in the "free development of individualities" (705).

37. Cockrel in *Leviathan* (1970), 36.

38. Cockrel in *Leviathan* (1970), 38.

39. *Brief History of the League,* in LRBW, Internal and Organizational Materials, Box 1, Folder 19, Detroit Revolutionary Movements Records, LR000874, Walter P. Reuther Library.

40. A document entitled "The Black and Poor People's Scientific Research Organization" points to some of the ways that this mobilization might happen, from hiring Black geographers to raising funds to purchase a computer. The document emphasizes that technical and scientific workers in the organization must always be secondary to, and respectful of, the "mass base of super-exploited workers." Black and Poor People's Scientific Research Organization, Box 1, Folder 34, Detroit Revolutionary Movements Records, LR000874, Walter P. Reuther Library.

41. Cockrel in *Leviathan* (1970), 36–37.
42. Cockrel in *Leviathan* (1970), 32.
43. Watson, "Black Editor."
44. On production and distribution, see Georgakas and Surkin, *Detroit,* chapter entitled "Finally Got the News."
45. See Georgakas, "Finally Got the News," 4.
46. Morgan Adamson, "Labor, Finance, and Counterrevolution: *Finally Got the News* at the End of the Short American Century," *South Atlantic Quarterly* 111, no. 4 (2012): 810–11. See also an account of the film in relation to other works: Morgan Adamson, *Enduring Images: A Future History of New Left Cinema* (University of Minnesota Press, 2018).
47. Robé, "Detroit Rising."
48. "The History and Derivation of the League of Revolutionary Black Workers," in *The Black Power Movement, Part 4: The League of Revolutionary Black Workers, 1965–1976* (LexisNexis, 2004), 2, www.lexisnexis.com/documents/academic/upa_cis/100390_BlackPowerPt4LRBW.pdf.
49. "History and Derivation," 2.
50. "History and Derivation," 4.
51. "History and Derivation," 4.
52. Watson, "To the Point of Production," 7.
53. Jonathan Beller, *The Cinematic Mode of Production: Attention Economy and the Society of the Spectacle* (Dartmouth College Press, 2006), 9.
54. Early Soviet cinema, notably Dziga Vertov's *Man with a Movie Camera,* was especially successful at showing how this came to be. Beller writes: "Through the rationalization, routinization, and standardization of certain aspects of industrial production, montage achieves new orders of particularity and expressivity in the visual. Montage as fragmentation and montage as the connecting of fragments are at once the condition of modern life and the condition for the production of meaning in modern life" (39).
55. Beller, 39.
56. Geschwender tends to refer to Tripp as the author in similar circumstances. Whether this is because he knew that Tripp wrote these pages or because he used that terminology in a more quotidian sense is unclear, as his concerns lay elsewhere.
57. For issues of *South End,* see *The Southend,* 1969 (newspaper), Box 17, Folder 15, Detroit Revolutionary Movements Records, LR000874, Walter P. Reuther Library.
58. Georgakas, "Finally Got the News," 4–5.
59. I have been unable to find the name of the woman who narrates this section in the existing secondary literature on *Finally Got the News.*
60. *Leviathan* (1970), 38. Cockrel's discussion in "Our Thing is DRUM!" is a nearly identical, but more expansive, version of a speech he delivers midway through *Finally Got the News.* It is also similar to Nancy Fraser's sense of *Cannibal Capitalism.* However, in contrast, as I argue in this chapter, the League's engagement with reproduction and gendered labor is limited.
61. See, e.g., *DRUM'S Program* in DRUM; Program, Box 1, Folder 1, Detroit Revolutionary Movements Records, LR000874, Walter P. Reuther Library;

Leviathan (1970); The League of Revolutionary Black Workers General Program, Box 1, Folder 86, American Left Ephemera Collection, Archives Service Center, University of Pittsburgh, Pittsburgh, Pennsylvania.

62. The limited extent to which they discuss reproduction and the family is evidenced, for example, in "History and Derivation," where they write in brief that "another facet of the working woman is that she physically produces a labor force" (32). However, although they also discuss the struggles of Black women in the labor market, they have little else to say on the subject.

63. Cockrel, *League of Revolutionary Black Workers on Repression Speech,* 12.

64. *Leviathan* (1970), 24; Louis Althusser, *On the Reproduction of Capitalism: Ideology and Ideological State Apparatuses,* trans. G. M. Goshgarian (Verso, 2014).

65. *Leviathan* (1970), 21.

66. Another example of this is that the League sought to use Michigan's education decentralization act of 1969 (Public Act 244) as a way to organize the community as a step toward greater control of education and ultimately state power, even though "community control of like anything is not the solution," and despite the objections of other radical groups that involvement with it was always already compromised. *Leviathan* (1970), 26–27.

67. Georgakas and Surkin, *Detroit,* 45.

68. "Black Student Voice," in *Black Power Movement, Part 4.*

69. "Revolutionary Black Culture as Weapon," in *Black Power Movement, Part 4.*

70. Geschwender cites a brief mention of Fanon by Watson in the context of discussions about the Black Panthers at a talk "Perspectives: A Summary Session," the final presentation of the first year of the "Control, Conflict, and Change Book Club," June 8, 1971. Geschwender, *Class, Race, and Worker Insurgency,* 141.

71. Turner, *From Counterculture to Cyberculture,* 17.

72. Robinson, *Black Marxism,* 170–71.

6. The Black Panthers

1. *The Black Panther: Intercommunal News Service,* June 26, 1971, 3.

2. Robyn C. Spencer, *The Revolution Has Come: Black Power, Gender, and the Black Panther Party in Oakland* (Duke University Press, 2016), 31.

3. "What We Want Now! What We Believe," *The Black Panther: Black Community News Service,* May 15, 1967, 3; "Black Panther Party Program, March 29, 1972, Platform," *The Black Panther: Intercommunal News Service,* May 13, 1972, B.

4. Huey P. Newton, "The Technology Question," in *The Huey P. Newton Reader,* ed. David Hilliard and Donald Weise (Seven Stories Press, 2019).

5. See the first issue of *The Black Panther: Black Community News Service* from April 25, 1967, which is focused on the Denzil Dowell case. On the significance of the Dowell case for the Panthers, see Joshua Bloom and Waldo Martin Jr., *Black Against Empire: The History and Politics of the Black Panther Party* (University of California Press, 2016), esp. "Preface to the 2016 Edition" and chap. 2, "Policing Police."

6. Bobby Seale, *Seize the Time* (Random House, 1970), 140–41.

7. Huey P. Newton, "The War against the Panthers: A Study of Repression in America" (PhD diss., University of California, Santa Cruz, 1980).

8. "Computers to Watch Riots," *The Black Panther: Black Community News Service,* May 18, 1968, 3. On the politics of race and the development of computer programming from a different, although relevant, angle, see Tara McPherson, "U.S. Operating Systems at Mid-Century: The Intertwining of Race and UNIX," in *Race After the Internet,* ed. Lisa Nakamura and Peter Chow-White (Routledge, 2011).

9. Newton, *Newton Reader,* 342–46.

10. Newton, 346.

11. Newton and Erikson, *In Search,* 23.

12. Karl Marx and Frederick Engels, *Manifesto of the Communist Party,* trans. Samuel Moore, February 1848, 20, www.marxists.org/archive/marx/works/download/pdf/Manifesto.pdf.

13. Jackie Wang, *Carceral Capitalism* (semiotext(e), 2018), 64.

14. Chris Booker, "Lumpenization: A Critical Error of the Black Panther Party," in *The Black Panther Party (Reconsidered),* ed. Charles E. Jones (Black Classic Press, 2005).

15. Seale, *Seize the Time,* 30.

16. Alongside Booker and Wang, see Laura Pulido, *Black, Brown, Yellow, and Left: Radical Activism in Los Angeles* (University of California Press, 2006); and Clyde Barrow, *The Dangerous Class: The Concept of the Lumpenproletariat* (University of Michigan Press, 2020).

17. Eldridge Cleaver, "On the Ideology of the Black Panther Party," *The Black Panther: Black Community News Service,* June 6, 1970, insert.

18. Larry Jones, "Up from Capitalism," *The Black Panther: Black Community News Service,* May 19, 1969, 16.

19. Charles Bursey, "Unite," *The Black Panther: Black Community News Service,* May 19, 1969, 17.

20. Lumpen vanguard politics were, however, an important part of Elaine Brown's indoctrination into the Los Angeles chapter by John and Ericka Huggins. Elaine Brown, *A Taste of Power: A Black Woman's Story* (Anchor Books, 1994), chap. 7, "Living for the People."

21. Newton, *Newton Reader,* 178.

22. Cedric Johnson, "Huey P. Newton and the Last Days of the Black Colony," in *African American Political Thought: A Collected History,* ed. Melvin L. Rogers and Jack Turner (University of Chicago Press, 2021), 631–59.

23. See Huey P. Newton, "On the Defection of Eldridge Cleaver from the Black Panther Party and the Defection of the Black Panther Party from the Black Community," in *Newton Reader;* and *Black Against Empire,* chap. 15, "Rupture."

24. For an analysis of this passage that connects to chapter 1's discussion of capitalism and the Black colony, see John W. Elrick, "Simulating Renewal; Postwar Technopolitics and Technological Urbanism," *Environment and Planning D: Society and Space* 38, no. 6 (2020).

25. Étienne Balibar, "In Search of the Proletariat: The Notion of Class Poli-

tics in Marx," in *Masses, Classes, and Ideas,* trans. James Swenson (Routledge, 1994), 131.

26. Balibar, 144; Balibar, *The Philosophy of Marx,* trans. Chris Turner (Verso, 2007), 39–54.

27. Elaine Brown and Angela Davis, "Welcome Home, Angela Davis," *The Black Panther: Intercommunal News Service,* March 4, 1972, F-G.

28. George L. Jackson, *Blood in My Eye* (Random House, 1972), 124.

29. Jackson, 65.

30. George Jackson, *Soledad Brother* (Lawrence Hill Books, 1994), 242.

31. Jackson, 251–52.

32. Jackson, 252.

33. Jackson, *Blood in My Eye,* 252, 7.

34. Davis, *Women, Race, and Class,* 223.

35. Angela Davis, "Women and Capitalism: Dialectics of Oppression and Liberation," in *The Angela Y. Davis Reader,* ed. Joy James (Blackwell, 1998), 163.

36. Davis, 173.

37. Davis, *Women, Race, and Class,* 5.

38. Davis, 11.

39. Saidiya Hartman, "The Belly of the World: A Note on Black Women's Labors," *Souls* 18, no. 1 (2016).

40. See Vladimir Ilyich Lenin, *Conspectus of Hegel's Book "The Science of Logic,"* trans. Clemence Dutt, www.marxists.org/archive/lenin/works/1914/cons-logic/index.htm; and C. L. R. James, *Notes on Dialectics: Hegel, Marx, Lenin* (Lawrence Hill, 1980).

41. Newton and Erikson, *In Search,* 25.

42. Huey P. Newton, "A Statement by Huey Newton . . . ," *Black Panther Intercommunal News Service,* March 13, 1971, 2.

43. Huey P. Newton, "Speech Delivered at Boston College: November 18, 1970," in *To Die for the People: The Writings of Huey P. Newton* (Vintage, 1972), 27–28.

44. John Narayan, "Huey P. Newton's Intercommunalism: An Unacknowledged Theory of Empire," *Theory, Culture, and Society* 36, no. 3 (2017); Delio Vasquez, "Intercommunalism: The Late Theorizations of Huey P. Newton, 'Chief Theoritician' of the Black Panther Party," *Viewpoint Magazine,* June 11, 2018, viewpointmag.com/2018/06/11/intercommunalism-the-late-theorizations-of-huey-p-newton-chief-theoretician-of-the-black-panther-party/.

45. Newton, *Newton Reader,* 185, 271.

46. Although politically very different, there are resonances here with Martin Heidegger, "The Question Concerning Technology" (1954), in *The Question Concerning Technology and Other Essays* (Harper Perennial, 2013).

47. Huey Newton, "Intercommunalism" (1974), *Viewpoint Magazine,* June 11, 2018, viewpointmag.com/2018/06/11/intercommunalism-1974/.

48. Newton, *Newton Reader,* 249.

49. Michael Hardt and Antonio Negri, *Empire* (Harvard University Press, 2000).

50. Wendy Brown, *Walled States, Waning Sovereignty* (Zone Books, 2010), 22–23.

51. This position led him to disagree with some movements that he actually supported. For example, he believed that if the Republic of New Afrika set up a Black nation in the South, it would become subject to the capitalist imperialism of the United States, just like other countries in the Third World. Huey Newton, "To the Republic of New Africa," in *Off the Pigs! The History and Literature of the Black Panther Party,* ed. G. Louis Heath (Scarecrow Press, 1976).

52. Newton, *Newton Reader,* 248.

53. Newton, 188.

54. Newton and Erikson, *In Search,* 30; Newton, *Newton Reader,* 184.

55. Eldridge Cleaver, "Community Imperialism," *The Black Panther: Black Community News Service,* May 18, 1968, 10.

56. Miranda Joseph, *Against the Romance of Community* (University of Minnesota Press, 2002).

57. Narayan, "Huey P. Newton's Intercommunalism," 18; Newton and Erikson, *In Search,* 33–35.

58. Lee Lockwood, *Conversation with Eldridge Cleaver, Algiers* (McGraw Hill, 1970), 63.

59. Brown, *Taste of Power,* 276–82.

60. Alondra Nelson, *Body and Soul: The Black Panther Party and the Fight against Medical Discrimination* (University of Minnesota Press, 2011), 11–12. See also David Hilliard, ed., *The Black Panther Party: Service to the People Programs* (University of New Mexico Press, 2008). Hilliard draws from Black Panther Party, *CoEvolution Quarterly* 3 (1974).

61. Spencer, *Revolution Has Come,* 4.

62. Spencer, 4.

63. Newton, *To Die for the People,* 104.

64. Newton, *Newton Reader,* 279.

65. Newton, *To Die for the People,* 104.

66. Newton, *Newton Reader,* 328.

67. Newton, 328.

68. Newton and Erikson, *In Search,* 76.

69. Newton, *Newton Reader,* 277–78.

70. Newton, 320. Although different in many respects, my approach to Newton's technology question and intercommunalism has been inspired by McKenzie Wark, *Molecular Red* (Verso, 2015).

71. "Ericka C. Huggins Oral History Interview Conducted by David P. Cline in Oakland, California, 2016 June 30," video, Civil Rights History Project, American Folklife Center, Library of Congress Archive of Folk Culture, www.loc.gov/item/2016655435/.

72. Black Panther Party, *CoEvolution Quarterly* 3 (1974): 10.

73. Michel Foucault, *The Order of Things* (Routledge, 2005), xvi.

74. Newton, *Newton Reader,* 273.

75. On the Panthers' constitutionalism generally, and especially the extent that they affirmed the U.S. Constitution with the Ten Point Program, as

well as their participation in electoral campaigns and lawsuits during the 1970s, see Devin Fergis, "The Black Panther Party in the Disunited States of America: Constitutionalism, Watergate, and the Closing of the Americanists' Minds," in *Liberated Territory: Untold Local Perspective on the Black Panther Party,* ed. Yohuru Williams and Jama Lazerow (Duke University Press, 2008).

76. For contemporary coverage, see "Panthers Seeking New Constitution," *New York Times,* June 20, 1970; and Paul Delaney, "Panthers to Reconvene in Capital to Ratify Their Constitution," *New York Times,* September 8, 1970.

77. George Katsiaficas, "Organization and Movement: The Case of the Black Panther Party and the Revolutionary People's Constitutional Convention of 1970," in *Liberation, Imagination and the Black Panther Party: A New Look at the Black Panthers and their Legacy,* ed. Kathleen Cleaver and George Katsiaficas (Routledge, 2001).

78. The original document is available on George Katsiaficas's website at www.eroseffect.com/articles/Blog%20Post%20Title%20One-ekxdd (accessed December 22, 2023).

79. Katsiaficas, "Organization and Movement," 154.

80. Huey P. Newton, "Towards a New Constitution," *Black Panther: Black Community News Service,* June 13, 1970, insert.

81. Simon Sadler, "Mandalas or Raised Fists? Hippie Holism, Panther Totality, and Another Modernism," in Blauvelt, *Hippie Modernism.*

82. Newton, *Newton Reader,* 39.

83. Katsiaficas, "Organization and Movement," 154.

84. Newton, *Newton Reader,* 43.

85. Newton, 42.

86. Newton, 43.

87. *The Black Panther: Black Community News Service,* March 13, 1971, 2.

Index

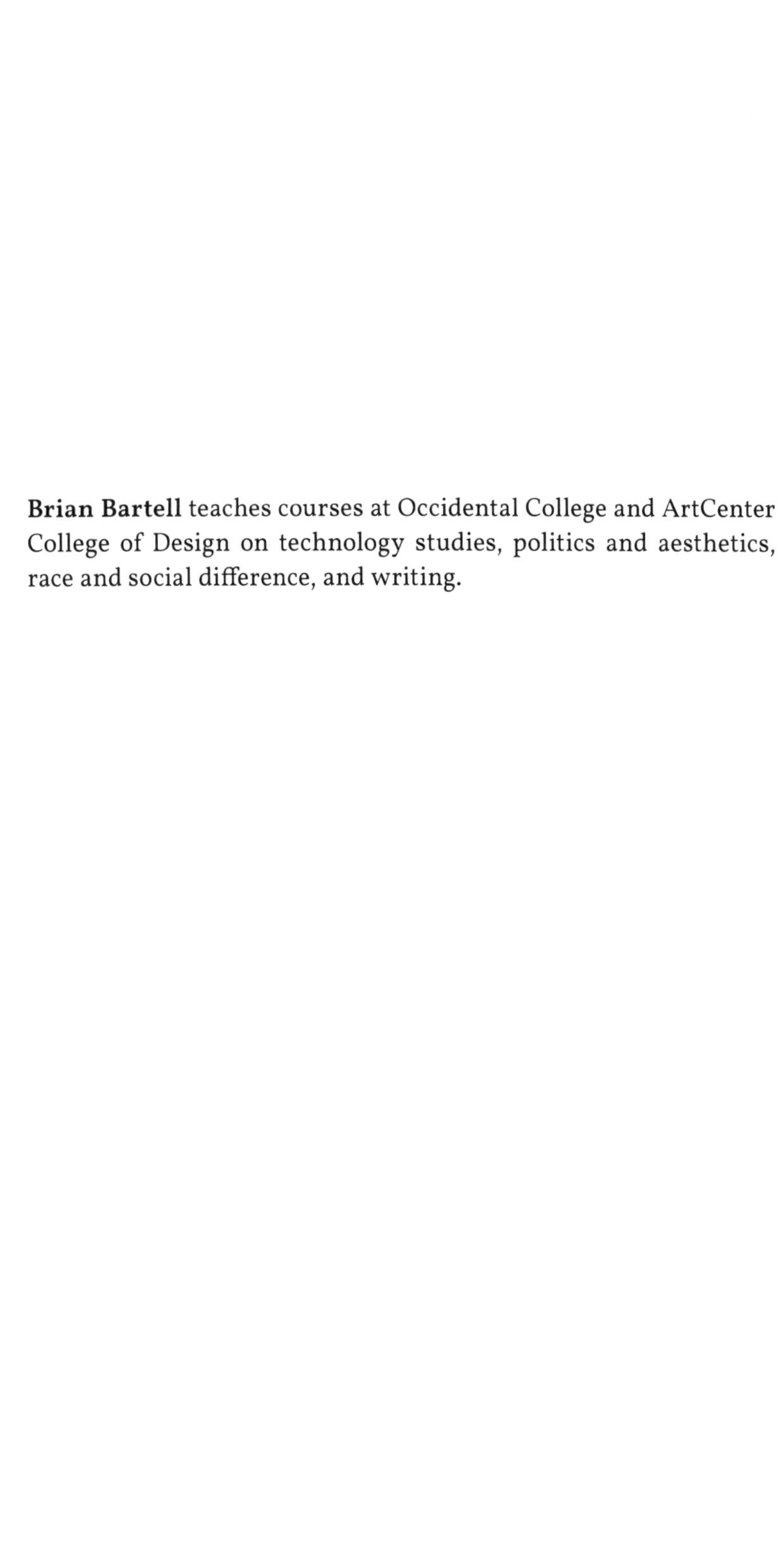

Brian Bartell teaches courses at Occidental College and ArtCenter College of Design on technology studies, politics and aesthetics, race and social difference, and writing.